健康生活系列

程安琪 著

A Chicken in Every Pot

家家锅中有只鸡

中国农业大学出版社

作者的话

从小我就爱吃鸡，因为妈妈在家中教做菜，让我吃到许多好吃的鸡肉料理，无论蒸、煮、炒、炸、红烧我都爱吃，而记忆中最好吃的是"山东烧鸡"和"香酥鸡"的鸡骨头，因为学生吃了鸡肉，拿到后面给我们解馋的是带有一点肉的骨头，但是花椒的香气渗入鸡骨中，真是连骨头都好吃！

鸡因为体型不大，可以整只烹调后各取所爱的部分，像我家煮一锅鸡汤，女儿爱鸡腿、儿子吃鸡胸、婆婆就选肉少的鸡爪和鸡翅；鸡也常被分割成不同的部位出售，可以按照个人的喜好选购，只买鸡腿、鸡翅或鸡胸来做不同的菜式，鸡肉实在是家常最方便烹调的一种肉类食品。

鸡肉不但适合做家常菜，它和鸭子、鸽子同属家禽类，是传统中国酒席上六类大菜中不可缺少的一类，因此像"去骨八宝鸡"、"富贵烤鸡"、"石榴鸡"、"东江盐焗鸡"一类的功夫菜，就可以端上请客的台面。鸡肉不似牛肉、羊肉有较重的气味，因此它适合许多种类的烹调方法，也适合和味浓或清爽的食材来做搭配，可以说浓妆淡抹两相宜，是非常好变化的一种食材。

虽说"鸡"是一种食材，但是因为鸡的品种很多，土鸡和肉鸡的口感截然不同，各有适合烹调的方式，可以做出不同的变化；而一只鸡的鸡胸和鸡腿就有不一样的肉质口感，各有适合的烹调方式，因此虽是一种食材，却能做出上百道的口味呢！

一个让我印象最深刻的鸡料理，这次我也把它收录在后面的食谱中，那就是"精华补身鸡汁"，记得妈妈因为心包膜钙化而做心脏手术之后，婆婆每天都买一只土鸡，蒸一碗原汁给她喝，10只鸡的精华喝下来，妈妈的精神和体力都恢复得很快，难怪我国自古即把鸡当做养生的补品，不但如此，它还是祭天拜祖的三牲之一，可见其重要性。

中国人是好客的，在农业社会、物资不丰富的年代，远方的朋友来访，会杀一只鸡待客。过去只有在年节时的餐桌上才会出现全鸡来打牙祭；原美国总统胡佛在竞选时才会以"家家锅中有只鸡"为竞选口号，希望每一个家庭的锅里都能常常有只鸡，以代表民生富裕。现在的社会当然不一样了，鸡已经成为大众化的食物，以现代化的技术来养鸡，鸡也有

自己的品牌，以先进的技术、垂直整合的经营、土鸡溯源系统、CAS、HACCP和ISO的把关，确保鸡肉来源安全无虞，使我们吃的安心、吃的开心！

一直以来我就想出版一本有关鸡的食谱，但是能用鸡做的菜实在是太多了，该以什么方式来呈现，才能使读者朋友完全受用呢？我想了很久，最后仍决定以鸡的部位来作划分。诚如我前面说过，鸡胸和鸡腿的肉质不同，它的处理方法及适合的菜式就不同，当你会处理鸡胸肉了，无论切成丝、柳、片、丁、粒、块，或是搭配任何蔬菜，就都难不倒你啦！自己变化用不同的蔬菜或加上不同的调味料，就是一道新菜了！所以在每一个篇章之处，我把处理方法和烹调重点特别整理出来，希望读者能先阅读一遍，知道该如何处理它，才能得到最好的结果，做出最好吃的菜肴。

另外，在全鸡的部分，我特别再分出白煮鸡和白煮鸡的一些变化吃法，鸡肉高蛋白质和低脂肪的特性，以白煮的方式更能完整呈现，但以现在小家庭的量，恐怕难以一次吃完，所以特别提供一些变化吃法给大家参考，在相关篇章中我也提到，买整只鸡要比分着买便宜、变化也多。

曾经有一本书《家家锅里有只鸡》是叙述大成集团的成长，我无法和他们相比，但是我很喜欢这个书名的意境，征得韩家寅先生的同意，我选了类似的书名，希望书中的102道鸡肉美馔，丰富每个家庭的餐桌，使每个厨房都飘散着鸡的香气！

I like chicken since I was a kid. When my mother was teaching cooking at home, I tasted many delicious chicken dishes and I enjoyed them all! I still remember the succulent taste of the bones of Chicken Salad with Cucumber and Crispy Chicken with Brown Peppercorn Flavored. Because the meat part was for the students to taste, my siblings and I only got the leftover bones and a little bit of meat as snacks. Although there was not much meat left, the fragrance of brown pepper corn had penetrated through the bone and I can still remember the taste!

When chicken cooked in whole, everyone can select his or her favorite part of a chicken. Take my family for example, my daughter likes chicken legs, my son likes the breasts, and my mother-in-law always picks the feet and wings. You may also choose to purchase specific parts of chicken in the market to make different dishes. The versatility of chicken makes it the most convenient meat for home cooking.

Chicken, along with duck and pigeon, is a type of poultry meat. Chicken is not only commonly used in home cooking but is also a key ingredient for banquet dishes. Compared to beef or lamb, chicken has a more subtle flavor. Therefore many different cooking methods are applicable to chicken. Chicken goes well with both subtle ingredients and intense seasonings because it does not overpower nor interfere with other flavors. You can be very creative with how to cook the chicken.

Although it is common to think of "chicken" as one ingredient, different breeds of chicken contain different textures. Therefore, one cooking method may be more suitable for one type of chicken than another. Furthermore, chicken breasts have different characteristics from chicken legs, and hence they should be handled differently.

Chicken is very nutritious. When my mother was recovering from her heart surgery, my mother-in-law would bring her a bowl of steamed, wholesome chicken broth made with one whole native free-range chicken everyday. After 10 days, my mother had become a lot more energetic and her physical condition had improved significantly. No wonder chicken is considered to be a life prolonging ingredient since ancient Chinese culture. The recipe of this chicken stock, Concentrate Chicken Broth, is also included in this book.

Chinese are known for our hospitality. At the time when materials were less abundant, putting a chicken dish on the table was only for special occasions such as hosting an old friend from far nway or celebrating

the Chinese New Year.In 1928, the campaign slogan of United States President Hoover was "A Chicken in Every Pot", which symbolized a wealthy in the late 20s. Nowadays chicken is a very common food. They are raised in farms using modern technologies, they have their own brands, their breeds are traceable, and they are under the inspection of regulations such as CAS, HACCP and ISO.

I've always wanted to write a cookbook about chicken. However, there are so many chicken recipes! How can I organize them so that it is most structured and reader-friendly? I decided to categorize the recipes based on different parts of chicken. As I mentioned earlier, different parts of chicken meat have different characteristics and therefore should be utilized differently. Once you know how to prepare chicken breast, the same procedure is applicable to chicken breast strings, slices, cubes and chunks. You can then create dishes with your choices of vegetables and seasonings. In the beginning of each chapter, I compile an introduction of the characteristics, preparation and cooking tips of the part of chicken featured in that chapter. I hope you would read the introduction first because it conveys the essential messages for perfecting the following recipes.Boiled chicken emphasizes the health benefit of chicken, which is high in protein and low in fat. Therefore, I designate a chapter (ma, is boiled chicken a chapter like "chicken breast", or it is just a special section in "whole chicken"？if it's only a section in "whole chicken", say "I designate a section in "whole chicken"") to show you how to cook boiled chicken creatively. It may be challenging for a small family to finish a whole boiled chicken in one meal, and these recipes will turn the leftovers into more delicious dishes.

There is a book, "A Chicken in Every Pot", which tells the success story of 大成集团. I really like the profoundness of the title, and therefore I've asked Mr. 韩家寅's permission to use a similar title for this cookbook. I hope the 102 recipes in this book will enrich your family meals and put a chicken in your pot!

作者的话

细说鸡——了解鸡，你才能做出好吃的鸡

All about Chicken - Understanding is the First Step

在我们用鸡肉做菜之前，要对鸡先有些基本的认识，在知道了不同品种的鸡的特性后，我们才能依照要做的菜式去选购适合的鸡。"鸡"，总体上分为生蛋的"蛋鸡"和吃肉的"肉鸡"，常见的食用肉鸡又分为土鸡、仿土鸡、放山鸡、乌骨鸡和白肉鸡。另外还有体积比较小的古早鸡或是玉米鸡、斗鸡、珍珠鸡，这些都是比较少见的鸡种了。至于有些人为了要求肉质的紧实，因而有阉鸡的产生，就是更专业的问题了。每一种鸡都有它的特色和适合烹调的方法，简单的可以区分为：

Before cooking chicken, you should understand the basics about chicken. In general there are two types of chicken: laying chicken and meat chicken. The meat chicken in Taiwan include native chicken, free range chicken, simulate native chicken, silky bantam (also called black meat chicken), and broiler. Other less common ones include old-type chicken, which is smaller in size, corn-feed chicken, fight chicken, and pearl chicken. Different types of chicken have different characteristics and therefore the suitable cooking methods may vary.

土鸡：

真正的纯种土鸡已经十分罕见了，目前常见的土鸡也曾经是经过许多次引种，选择优良的品种与当地的鸡种育种而成。在不同的地区有不同的品种和名称，例如内门鸡、竹崎鸡、草鸡、龙冈鸡、清远鸡、三黄鸡、九斤黄等都是。现在在超市常见的"鹿野嫩黄土鸡"是属于新育种成功的特殊鸡种。

一般而言，台湾地区的黑羽土鸡的体型较小、瘦长，肉质紧实、鲜美，母鸡1.5～1.8千克，公鸡在2.5～3千克，饲养时间为18～20周，多半用来炖汤、做白煮鸡、清蒸或红烧。而其他品种的土鸡在体型和外观上又有不同的特色，有的体型较大，脚胫也属于黄色，但仍具有黑羽土鸡的肉质特色。

Native chicken：

The real native chicken are very rare nowadays. There native chicken we are having now are copulation for many times. There are different breeds among native chicken, and they have different names in different places. "Dachan Mi Hon Mi San native chicken" is an example of new breed of native chicken.

In general they are smaller in size, typically ranging from 1.5 to 1.8 kg per hen, and 2.5 to 3 kg per cock. It takes about 18~20 weeks to raise it. It is tasty and flavorful. They are suitable for making

纯土鸡 Native chicken

鹿野嫩黄土鸡 Dachan Mi Hon Mi San native chicken

soup and stew, and they also taste good boiled or steamed.

仿土鸡：

仿土鸡又被称为半土鸡，经过许多次外来鸡种的融合，体型比纯土鸡大，为2～2.7千克，养成时间为13～14周。也具有鲜甜的肉质，适合蒸、炖煮、红烧、炒、烩等许多类型的烹调方法。其中母鸡和公鸡的肉质差异挺大的，母鸡的皮下油脂较多，但肉质较嫩，读者可以自行比较、选购。

Simulate native chicken：
Simulate native chicken are also called "half native chicken" because they are native chicken that have been crossbred multiple times with foreign breeds. They are larger than native chicken, typically ranging from 2 to 2.7 kg, and the breeding time is roughly 13 to 14 weeks. They are suitable for making soup, steaming, stewing, stir-frying and many other cooking methods. Notable difference in tenderness of meat can be found between the hen and cock.
Hens are more tender; however with higher fat content under the skin.

仿土鸡 Simulate native chicken

放山鸡：

土鸡和仿土鸡都有以林间山野放养的方式来饲养的，即为俗称的放山鸡，因为饲养的环境不同，肉质更为紧实、有弹性，体型也比较大，有3～4千克之重，以至5千克之重的大鸡。

Free—ranged chicken：
Free-range chicken are chicken that are raised with freedom to roam around in open space. Since they are provided with an environment to exercise, the texture of the meat is firmer than those that

are raised within confined space. Free-range chicken are larger in size, typically ranging from 3 to 5kg.

白肉鸡：
　　一般通称为饲料鸡、肉鸡，占有目前鸡肉市场约1/2的量，因为肉质软嫩，适合炸、烤、炒、熏、卤，市面上常见的鸡腿便当、炸鸡、鸡排堡、卤鸡，或是用鸡肉做的加工食品，大部分用的是白肉鸡。现在的白肉鸡因为育种的进步和饲料的营养配方有改进，因此只要36～38天便可以达到标准重量和良好的品质。

Broiler：
Usually we just call it meat chicken. It is the most common breed of chicken in the market nowadays. They are suitable for deep-frying, baking, stir-frying, smoking, and stewing due to the tenderness of the meat. Most chicken meat used in lunch boxes, fried-chicken, chicken sandwiches, and processed food products is this kind of chicken. It only takes about 36~38 days to breed broiler.

白肉鸡 Broiler

乌骨鸡：
　　乌骨鸡有一身柔软的白色羽毛，在中国人的传统观念中，乌骨鸡比较补，因此许多炖补的药膳是用乌骨鸡做的。乌骨鸡的脂肪含量比白肉鸡低，而蛋白质含量高又好消化、易吸收，也增加爱食乌骨鸡者的信心。

silky bantam：
Silky bantam are also called black meat chicken. This type of chicken is considered more nutritious in Chinese culture and is a key ingredient in many Chinese herbal healthy dishes. There are also medical reports showing that the meat of black meat chicken is basic, which can lead to better health and prolonged life.

乌骨鸡 Silky Bantam

鸡 Contents

细说全鸡 All about Whole Chicken

Part 1 全鸡 Whole chicken

- 18 · 豉油卤鸡 Braised Chicken
- 19 · 山药手撕卤鸡 Braised Chicken with Yam Salad
- 20 · 三杯鸡 Three Cups Chicken
- 22 · 富贵烤鸡 Baked Stuffed Chicken
- 23 · 成都子鸡 Stir-fried Chicken, Chengdu Style
- 24 · 金针云耳烧鸡 Stewed Chicken with Fungus & Lily Flower
- 26 · 鲜茄烧鸡 Stewed Chicken with Fresh Tomatoes
- 28 · 沙锅油豆腐鸡 Stewed Chicken with Fried Tofu in Casserole
- 30 · 香烤手扒鸡 Home Style Baked Chicken
- 31 · 清蒸瓜子鸡汤 Steamed Chicken Soup with Pickled Cucumber
- 32 · 补身精华鸡汁 Concentrate Chicken Broth
- 34 · 沙锅鲍鱼土鸡汤 Chicken Soup with Abalone in Casserole
- 35 · 麻油鸡／麻油鸡面线 Chicken Soup with Sesame Oil Fragrant and Noodle
- 36 · 凤梨苦瓜鸡 Chicken Soup with Bitter Gourd & Pickled Pineapple
- 37 · 韩式参鸡汤 Ginseng Chicken Soup, Korean Style
- 38 · 赤小豆山药煲鸡汤 Chicken Soup with Red Beans and Yam

Part II 白煮鸡及它的变化菜式　Uses of Boiled Chicken

40・白煮鸡　Boiled Chicken
41・五香盐水鸡　Salty Chicken
42・醉鸡　Wined Chicken
43・口水鸡　Mouth Watering Chicken Salad, Sichuan Style
44・蛰皮手撕鸡　Chicken and Jellyfish Salad
45・东安鸡　Stir-fried Chicken, Dongan Style
47・怪味鸡　Flavored Chicken Salad
48・棒棒鸡　Bon Bon Chicken
49・三蔬拌鸡丝　Chicken Salad with Vegetables
50・鸡丝凉面　Cold Noodles with Chicken
51・泰式凉拌鸡丝　Chicken Salad, Thai Style
52・马铃薯鸡肉沙拉　Chicken & Potato Salad
53・芥末鸡肉沙拉　Chicken Salad with Mustard Dressing
54・炖鸡煨面　Chicken Soup Noodles

细说鸡胸　All about chicken Breasts

61・香根银芽炒鸡丝　Stir-fried Chicken Shreds with Bean Sprouts
62・炒鸡丝拉皮　Stir-fried Chicken Shreds Salad
63・豌豆炒鸡丝　Chicken Shreds with Baby Snow Peas
64・鸡丝炒牛蒡　Stir-fried Chicken with Bunduck
66・雪菜百叶烩鸡丝　Chicken Shreds with Pickled Mustard Green
68・瑶柱鸡丝羹　Chicken Shreds with Dried Scallop Potage
70・柠檬鸡片　Fried Chicken with Lemon Sauce

71・芥汁鸡片 Pouched Chicken with Mustard Sauce
72・韩风炒鸡肉 Stir-fried Chicken, Korean Style
74・味噌酱拌鸡柳 Chicken Salad with Miso Sauce
76・鸿喜腐乳鸡 Stir-fried Chicken with Fermented Tofu Sauce
78・辣子鸡丁 Diced Chicken with Peppers
80・川香鸡片 Stir-fried Chicken with Bean Paste
82・墨西哥鸡肉法西达斯 Chicken Fajitas
84・咖喱鸡排 Fried Chicken with Curry Sauce
85・红莓鸡片 Deep Fried Chicken with Cranberry Sauce
86・番茄起司烤鸡胸 Baked Chicken with Tomato Sauce
88・凯萨鸡肉沙拉 Chicken Caesar Salad
90・酥炸鸡条佐双酱 Fried Chicken with Salsa and Tar-tar Sauce
92・生炒鸡松 Stir-fried Minced Chicken
94・杂菜鸡丁 Stir-fried Chicken with Vegetables
96・鸡茸鲍鱼羹 Minced Chicken with Abalone Potage
97・鸡茸烩瓜丝 Minced Chicken with Wax Gourd
98・碎米鸡丁 Diced Chicken with Peanuts
99・泰式辣炒鸡肉 Spicy Chicken, Thai Style
100・咸鱼鸡粒炒饭 Stir-fry Rice with Salty Salmon & Chicken

细说鸡腿 All about Chicken Legs

Part 1 去骨鸡腿 Boneless Chicken Legs

106・参杞醉鸡卷 Wined Chicken with Chinese Herbs
108・腰果鸡丁 Diced Chicken with Cashew
110・香芹九层鸡丁 Diced Chicken with Basil & Celery

- 112・宫保鸡丁　Chicken with Gung-Bao Sauce
- 113・左宗棠鸡　Stir-fried Chicken, Hunan Style
- 114・香蒜烹鸡块　Fried Chicken with Garlic Sauce
- 115・纸包鸡　Paper Wrapped Chicken
- 116・蚝油鸡丁蛋豆腐　Chicken and Egg Tofu in Oyster Sauce
- 117・沙茶鸡肉串　Bar-B-Q Chicken Skewers with Sha-cha Sauce
- 118・梅干菜蒸鸡球　Steamed Chicken with Fermented Cabbage
- 119・京都子鸡　Chicken with Jing-Du Sauce
- 120・香酥鸡排堡　Chicken Hamburgers
- 122・八宝鸡排饭　Steamed Chicken Rice Pudding
- 124・椒麻鸡　Chicken Salad with Spicy Sauce
- 126・红烩鸡腿排　Chicken Legs with Tomato Sauce
- 128・北菇滑鸡煲　Chicken with Mushrooms in Casserole
- 130・去骨盐酥鸡　Crispy Chicken with Spices
- 132・荷叶粉蒸白果鸡　Steamed Chicken Packed with Lotus Leaves
- 133・亲子井（两碗份）　Rice Cover with Chicken & Egg Sauce (2 Servings)
- 134・鸡肉丸子汤　Chicken Meatballs soup
- 135・干锅鸡　Stir-fried Chicken in Casserole
- 136・咸鱼鸡粒豆腐煲　Chicken with Salty Fish & Tofu in Casserole
- 138・沙茶鸡肉炒面（四人份）　Stir-fried Noodles with Chicken in Sha-Cha Sauce (4 Servings)

Part II 带骨鸡腿　Bone-in Chicken Legs

- 141・轻烟熏鸡腿　Smoked Chicken
- 142・山东烧鸡　Chicken Salad with Cucumber
- 143・葱油鸡腿　Steamed Chicken Legs with Green Onion
- 144・香酥鸡腿　Brown Peppercorn Flavored Crispy Chicken
- 145・酸辣木耳鸡　Hot and Sour Chicken with Fungus

146· 泰式咖喱鸡 Curry Chicken, Thai Style
147· 韩式辣煮鸡 Spicy Chicken, Korean Style
148· 意式鲜疏烤鸡腿 Baked Chicken & Vegetables, Italian Style
149· 蒜头焗鸡腿 Bake Chicken with Garlic
151· 琥珀鸡冻 Jellied Chicken
152· 洋葱烧鸡 Stewed Chicken with Onion
153· 梅酱鸡 Chicken With Plum Sauce
155· 红烧香菇竹笋鸡 Stewed Chicken with Mushroom & Bamboo Shoot
157· 炸鸡腿菜饭 Deep-freied Chicken with Vegetable Rice

细说鸡翅 All about Chicken Wings

159· 辣烤鸡翅 Spicy Wings
160· 龙凤串翅 Stuffed Chicken Wings with Ham
161· 杏鲍凤翼煲 Chicken Wings and Mushrooms in Casserole
162· 玉米烩鸡翅 Stewed Chicken Wings with Corn
163· 香辣凤翼 Spicy Chicken Wings
164· 笋烧双宝 Stewed Two Treasures with Baboo Shoot
165· 香柠焖鸡翅 Stewed Wings with Lemon

细说鸡爪 All about Chicken Feet

166· 北菇炖凤爪汤 Chicken Feet Soup with Shitake Mushrooms
167· 豉汁凤爪 Steamed Chicken Feet with Black Bean Sauce

细说全鸡
All about Whole Chicken

Part I 全鸡 Whole Chicken

在现代的家庭中，煮一整只全鸡的机会越来越少，但是煮整只鸡的滋味是不同的。整只鸡因为没有切口，鸡的原汁会留在皮与肉之间，同此鸡肉的鲜甜味不会流失到汤里面，比剁成块再煮的鸡肉要好吃，因此一般做白煮鸡、盐水鸡都是用全鸡来煮的。

即使把全鸡剁成块，因为带有骨头，用来炖煮、红烧时，肉香混合着骨香，味道会更浓、更美，从另一个角度来看，一只鸡之中，较贵的是鸡腿和鸡翅，买全鸡比买两只鸡腿贵一点，但是煮出来的效果好一倍，何乐不为？即使不是整只鸡一次煮，买全鸡也是比较划得来的。

It is less and less common to cook a whole chicken in modern families. However, the full flavor of whole chicken cannot be mimicked. Since there is no opening or cut when cooking a whole chicken, the juice and the succulent flavor are preserved inside of the chicken. The meat of a whole chicken is more flavorful, which is why whole chicken are normally used when making "Boiled Chicken" or "Salty Chicken".
Another reason to cook with whole chicken is the additional flavor coming from the bone when making a stew or soup. Furthermore, whole chicken only cost slightly more than two chicken legs. It is more economic to buy a whole chicken, even if you don't cook it all at once.

利用全鸡来做的菜，基本上是希望鸡的鲜美滋味浓一些，因此常会选用土鸡、仿土鸡或放山鸡，如果是用来煮汤的，就要选购成熟一些的鸡，汤中才有肉香；如果是红烧或烤，就买嫩一点的小鸡。

无论是煮汤或是红烧，诚如我在前言中说过，每一个品种的鸡的肉质不同、公鸡和母鸡口感不同、鸡的重量不同，都会对烹煮的时间产生差异；同时对鸡肉软烂度的喜好也因人而异，基本上，鸡块剁的越大，需要的时间越长；喜欢吃口感比较嫩的人，只烧20～25分钟，但也有人要烧1小时以上，把它烧烂；要吃带肉的鸡汤和只爱喝鸡汤，炖煮时间也差很多的。因此在后面食谱中我所写的"炖煮的时间"，只是给读者一个参考，在实际烹煮时要试一下，再做增减。

Native chicken, simulate native chicken or free-range chicken are often chosen for dishes using whole chicken because they are more flavorful. If making soups, more matured chickens are preferred; if making stews or baked dishes, younger chickens are more suitable. As I mentioned in preface, the cooking time of a chicken is decided based on the type, sex and size of the chicken, and also your personal preference of how tender you like the meat to be. Generally the larger the pieces are, the longer the cooking time is. If you like chewier textures, it only takes 20~25 minutes to cook the chicken pieces. However, some people will cook it for over an hour until the meat is falling off the bone. Therefore, the cooking time in the recipes is only my recommendation and you may modify it accordingly.

Part II 白煮鸡及它的变化菜式 Uses of Boiled Chicken

　　白煮鸡是最能保持鸡的原味的烹煮方法，尤其是肉质甜美的优质鸡肉，最能吃出鸡本身的鲜甜味。如何煮出好吃的白煮鸡也有许多诀窍，最重要是煮鸡的时间和火候，而鸡的大小和煮的时间就有密切关系了。另外就是煮鸡的容器，要先找一个适当大小的锅，煮鸡时，只要鸡能完全浸入水中即可，如果水太多就会使鸡的鲜甜味流失到水中，因此锅子的宽度要让鸡可以放进去即可（图1）。也有人喜欢蒸，蒸的时候鸡汁会滴出，鸡肉会比较紧实。喜欢吃鸡皮脆爽的，在煮好后要立刻浸入有冰块的水中（图2），使鸡皮收缩起来。在后面的食谱里，我先教大家如何做白煮鸡。

　　鸡肉的脂肪含量低，用白煮和蒸更是健康的烹调法，而由白煮鸡又能变化出许多不同的菜式，除了年节祭拜时用白煮的全鸡之外，日常我们也可以煮一只白煮鸡，把鸡腿和鸡胸取下（图3、图4），分别做成不同的菜，再把鸡骨架再加料熬煮成鸡汤，完全不浪费。

　　白煮鸡能变化出来的菜式很多，因此我特别分出一个篇章举出13道菜来给大家参考，其实不同的调味料和配料就能做出不同味道的菜，读者可以自己再尝试！

Boiling is the best cooking method to retain and taste the natural flavor of chicken, especially when high quality chicken are available. There are many tips for boiling chicken, but the most important one is the control of cooking time and temperature. How long you should boil is closely related to the size of the chicken. The pot used to boil chicken needs to be large enough so that the chicken is fully under water while boling (pic1); however, using overly large pot with excess water will wash off the flavor of chicken. You may prefer steaming instead of boiling. Steamed chicken meat has a firmer texture due to the loss of chicken juice during steaming. if you favor jello—like texture for the chicken skin, you may soak the chicken in ice water immediately after it's cooked to tighten the skin(pic 2).

Chicken meat is low in fat, both boiling and steaming are considered the healthiest cooking methods for chicken. Furthermore, boiled chicken can be used creatively in many dishes. You may boil a chicken, remove the legs and breasts(pic3&4), and make chicken soup with the rest of the chicken and some additional ingredients. Nothing is wasted in this chicken!

I designated a chapter to show you how to utilize boiled chicken in many dishes. I included 13 recipes in this chapter; however there are unlimited possibilities with various seasonings and ingredients!

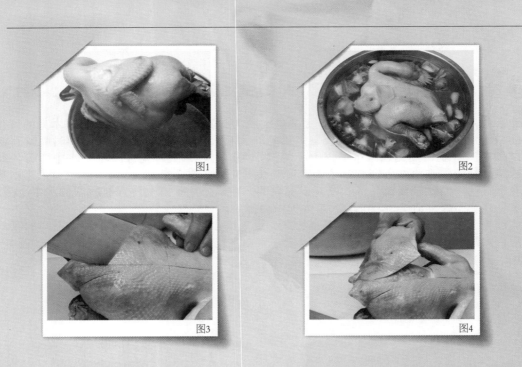

图1　图2　图3　图4

Braised Chicken

细说全鸡

豉油卤鸡
Braised Chicken

材料：
小的半土鸡1只（2～2.5千克）、五香包1包、麻油适量

调味料：
淡色酱油1 1/2 杯加深色酱油1/2杯、酒 1/2 杯、冰糖1～2大匙、盐 1/2 茶匙

做法：
1. 锅中将6杯水煮开后，放下五香包和调味料，以小火煮10分钟。
2. 鸡洗净，要把内部清洗干净，尤其是肺叶部分要清除，在滚水中烫约20秒钟，捞出、沥干水分。
3. 把鸡放入豉油汁中，盖上锅盖，以极小火泡煮约10分钟，翻面再煮约10分钟。
4. 关火，将鸡焖在卤汤中，约10分钟后取出鸡，待稍冷后，在表皮上涂少许麻油。
5. 食用时剁下所要的量，斩切成长块，也可以撕下鸡肉做凉拌菜使用。

※香港人称上好的酱油为豉油，有现成的瓶装出售，没有时可改用深、浅两色酱油，加糖等来煮制。五香包包括红葱头、大料、大蒜、葱和姜即可，香料味道不需太重。
※用过的卤汤可以冷冻保存，反复使用。对卤的食物有兴趣的读者可以参考本书作者的另一本食谱——《卤一卤&变一变》。

Ingredients:
1 small simulate native chicken (2 ~ 2.5 Kg.), 1 pack of five spice pack, sesame oil

Seasonings:
1 1/2 cups light colored soy sauce, 1/2 dark colored soy sauce, 1/2 cup wine, 1 ~ 2 tbsp rock sugar, 1/2 tsp salt

Procedures:
1. Put the seasonings and the five spice pack into a pot with 5 cups of boiling water, cook over low heat for 10 minutes.
2. Clean the chicken, especially the lung part. Blanch for 20 seconds, remove and drain dry.
3. Add chicken to the soy sauce broth, cover the lid and simmer for 10 minutes. Turn chicken over and simmer for another 10 minutes.
4. Turn off the heat. Soak chicken in broth for 10 minutes more. Remove and brush some sesame oil after the chicken cools a little.
5. Cut to pieces before serving or tear the meat to strings to make salad.

※The five spice pack including shallot, star anise, garlic, green onion and ginger. Wrap them in a piece of cheesecloth.
※You can frozen the soy sauce broth, just reheat it and cook with the ingredients you want to braised.

山药手撕卤鸡
Braised Chicken with Yam Salad

材料：
卤鸡 1/2 只、山药150克、
洋葱 1/3 个、香菜2根、炒过的芝麻1大匙

卤汁或淡色酱油2大匙、
醋或柠檬汁1大匙、糖 1/2 茶匙、
麻油1茶匙、蒜泥1茶匙、冷开水1大匙

做法：
1. 可参考前一道"豉油卤鸡"的方法做好卤鸡，待卤鸡凉后，可选用鸡腿或鸡胸肉部分，用手将鸡肉撕成粗条。
2. 洋葱切细丝，放入冰水中泡5～10分钟，去除辣气并增加脆度，沥干水分。
3. 山药削皮、切成细条或刨成丝，放在盘子上；香菜洗净、泡过水后切成段。
4. 调味料先调匀，一半淋在山药上。
5. 另一半调味料和鸡肉、洋葱和香菜拌匀，再放在山药上，撒上白芝麻即可。

Ingredients:
1/2 chicken, 150g. Chinese yam,
1/3 onion, 2 stalks cilantro,
1 tbsp fried sesame seeds

Seasonings:
2 tbsp light colored soy sauce or broth from braised chicken,
1 tbsp vinegar or lemon juice, 1/2 tsp sugar, 1 tsp sesame oil,
1 tsp mashed garlic, 1 tbsp drinking water

Procedures:
1. Refer to recipe on Page 13 to make the braised chicken. You may choose the chicken leg or breast meat, tear it into strips after it cools.
2. Shred onion, soak in ice water for 5～10 minutes to make it crispy. Drain.
3. Peel Chinese yam, cut into thin strips, then place on a serving plate; trim and soak cilantro in water, then cut to sections.
4. Mix seasonings in a bowl, pour half portion over yam.
5. Mix remaining sauce with chicken, onion and cilantro, put on top of yam. Sprinkle sesame seeds over the top. Serve cold.

细说全鸡

三杯鸡
Three Cups Chicken

材料：
仿土鸡1/2只（约1.2千克）、大蒜10粒、
老姜片10~12片、红辣椒1~2个、九层塔3~4枝

调味料：
黑麻油1/3杯、米酒1杯、酱油1/3杯、
冰糖1/2大匙、热水1杯

做法：
1. 鸡洗净，剁成约2厘米宽的块；大蒜小的不切、大粒的一切为二。
2. 锅烧热，放入麻油，加热至5分热时，放入姜片，以小火慢慢煎至香气透出。
3. 待姜片水分减少时，放入大蒜一起炒炸，至大蒜变黄时，放入鸡块，改成大火翻炒。
4. 炒至鸡肉变白，没有血水时，加入其余调味料煮滚，再倒入烧热的沙锅或铁锅中。盖上锅盖，用中小火煮至鸡肉熟透且水分收干，15~20分钟。
5. 放入切斜片的红辣椒拌炒一下，再放入九层塔叶，拌一下即可。

※现在已将传统的麻油、酱油和米酒各一杯的三杯鸡加以调整，以减少油分和咸度。
※九层塔在台湾地区属于广泛使用的香料。

Ingredients:
1/2 simulate native chicken (about 1.2Kg.), 10 cloves garlic, 10~12 slices ginger, 1~2 red chilies, 3~4 stalks basil

Ingredients:
1/3 cup black sesame oil, 1 cup wine, 1/3 cup soy sauce, 1/2 tbsp rock sugar, 1 cup hot water

Procedures:
1. Clean the chicken and cut into 2cm wide pieces. Halve the garlic if it is large.
2. Heat sesame oil in a wok until 100℃, fried ginger until brown and fragrant.
3. Add garlic in, when garlic turn golden, add chicken and stir-fry over high heat until the color of chicken have changed.
4. Add other seasonings, bring to a boil. Remove to a heated casserole or a heavy iron pot. Cover the lid and cook over medium~low heat for about 15~20 minutes until chicken is done and the sauce is absorbed.
5. Add red chili sections, stir evenly. Add basil leaves at last, mix and serve hot.

※Traditionally this dish is made with 1 cup of sesame oil, 1 cup of soy sauce and 1 cup of wine. This recipe has been modified and the amounts of oil and soy sauce have been reduced due to the health concern.

Baked Stuffed Chicken

※在烤的时候要将铝箔纸包转动数次，以使汤汁流动，鸡肉才入味。
※这道菜由江浙菜的"叫化鸡"改变而来，原本要包裹泥土来烤，喜欢荷叶香气的，也可以将干荷叶泡软刷干净，包着鸡来烤。家常做也可以改用蒸的。

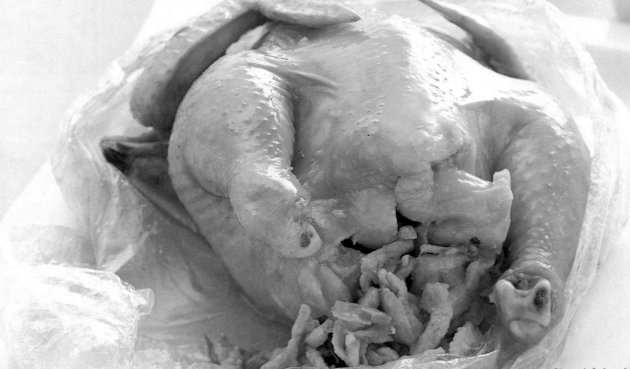

※Rotate the aluminum foil packet several times during baking so that the liquid inside can circulate and flavor the chicken.
※This dish is a modified version of "beggar's chicken", where the chicken is packed and baked in mud. If you like the fragrance of lotus leaves, you may wrap and bake the chicken with soaked soft and cleaned lotus leaves. This dish can also be steamed.

细说全鸡

富贵烤鸡
Baked Stuffed Chicken

材料：
小鸡1只（1.5～1.8千克）、肉丝75克、葱丝1杯、
冬菜或福菜 1/2 杯、大蒜4～5粒、玻璃纸1张、铝箔纸1大张、牙签2～3支

调味料：
（1）酱油3大匙、酒2大匙、胡椒粉 1/4 茶匙
（2）酒1大匙、酱油 1/2 大匙、糖 1/2 茶匙、盐 1/4 茶匙

做法：
1. 鸡的腹腔要特别清洗干净，去除肺脏等，再灌入滚水冲洗一下，擦干水分。
2. 鸡胸和鸡腿肉较厚之处，用叉子叉几下，再将胸骨压扁，内外均涂上调味料（1）腌20分钟。
3. 大蒜切厚片；冬菜（用白菜或芥菜叶做成的干菜）在水中泡5分钟，沥干备用。
4. 起油锅用2大匙油炒香大蒜片，放入肉丝和葱丝，炒至香气透出后加入冬菜和调味料（2）炒匀，盛出，装入鸡的肚子里，用牙签封口。
5. 玻璃纸上涂上油，包住鸡身（图1），外面再加包一层铝箔纸（图2，最好使用双层）。
6. 烤箱预热至220℃，放入铝箔包，烤约3个小时，至鸡肉已够烂为止。
7. 打开纸包，可以附馒头片或土司面包一起上桌夹食。

Ingredients:
1 small chicken (1.5～1.8kg.), 75g. pork strings,
1 cup shredded green onion, 1/2 cup fermented Chinese cabbage,
4～5 cloves garlic, 1 piece cellophane paper, 1 piece foil, 2～3 toothpicks

Seasonings:
(1) 3 tbsp soy sauce, 2 tbsp wine, 1/4 tsp pepper
(2) 1 tbsp wine, 1/2 tbsp soy sauce, 1/2 tsp sugar, 1/4 tsp salt

Procedures:
1. Clean chicken thoroughly, especially the belly cavity, rinse with hot water, and then wipe dry.
2. Pierce the thick part of chicken with a fork, such as legs and breast. Press down the breast to make it flat. Rub seasonings (1) all over the chicken, set aside for 20 minutes.
3. Slice garlic; soak fermented cabbage for 5 minutes, drain.
4. Heat 2 tablespoons of oil to sauté garlic, pork strings, and green onion. When fragrant, add fermented cabbage and seasonings (2), mix evenly, remove into the cavity of chicken belly. Seal with toothpicks.
5. Rub oil on cellophane pepper, wrap chicken in (pic 1), then wrap with foil (it's better to double the foil as the pic 2).
6. Preheat oven to 220℃, bake chicken for 3 hours until chicken is tender enough.
7. Unpack the chicken, serve with some steamed buns or bread.

成都子鸡
Stir-fried Chicken, Chengdu Style

材料：
嫩鸡半只（约600克）、红辣椒丁1大匙、葱末2大匙、
姜末1/2大匙、蒜末1/2大匙、芹菜末2大匙、花椒粒1茶匙

调味料：
辣豆瓣酱1大匙、盐1/2茶匙、糖1茶匙、镇江醋1/2大匙、
酒1大匙、淀粉水2茶匙

做法：
1. 鸡连骨带皮斩剁成2厘米的小块。
2. 炒锅内将3大匙油烧至极热，倒入鸡块大火炒拌1～2分钟。
3. 见鸡肉已半熟时，放进花椒粒再同炒，并将红辣椒与辣豆瓣酱落锅，继续再炒约2分钟。
4. 淋下酒，放下姜、蒜末，加糖、盐和醋调味，注入约1杯的热水，盖上锅盖，焖煮4～5分钟，至鸡块熟透为止。
5. 用淀粉水勾芡，炒拌均匀，再将芹菜末与葱末撒下，拌炒一下即可盛出。

※这道四川名菜应使用取自泡菜卤中的泡红椒，炒出来的鸡肉酸香兼具。

Ingredients:
1/2 chicken (about 600g.), 1 tbsp chopped red chili, 2 tbsp chopped green onion, 1/2 tbsp chopped ginger, 1/2 tbsp chopped garlic, 2 tbsp chopped celery, 1 tsp brown peppercorn

Seasonings:
1 tbsp hot bean paste, 1/2 tsp salt, 1 tsp sugar, 1/2 tbsp brown vinegar, 1 tbsp wine, 2 tsp cornstarch paste

Procedures:
1. Chop the chicken into 2cm pieces.
2. Heat 3 tablespoons of oil to stir-fry chicken over high heat for about 1~2 minutes.
3. Add brown peppercorn in, stir-fry for a while. Add red chili and hot bean paste, continue to stir-fry for 2 more minutes.
4. Pour wine in, then add ginger, garlic, sugar, salt, vinegar and 1 cup of hot water. Cover and simmer for about 4~5 minutes until chicken is cooked.
5. Thicken with cornstarch paste. Add chopped celery and green onion, mix well and serve.

※Chengdu is a big city in Sichuan Province of China. We should use the pickled red chilies to stir-fry this dish if we can get, it will be more fragrant and delicious.

金针云耳烧鸡
Stewed Chicken with Fungus & Lily Flower

材料：
半土鸡半只（约1.2千克）、干木耳1大匙、黄花菜30根、笋1棵、葱1根、姜3～4片

调味料：
酱油3大匙、酒1大匙、糖1茶匙、盐 1/3 茶匙、淀粉水少许

做法：
1. 鸡剁成块，放入滚水中烫20～30秒钟，捞出、洗净。
2. 木耳和黄花菜分别用水泡软，木耳要摘蒂、洗净，分成小朵；笋切片。
3. 用2大匙油先将葱、姜煎香，加入笋片炒过，先淋下酱油和酒，再加糖、盐及热水2杯，煮滚后把鸡和木耳放入拌合。
4. 大火煮滚后改成小火，盖上锅盖，烧煮约40分钟后放下黄花菜，再煮约20分钟至喜爱的烂度，试一下味道。
5. 如果汤汁仍多，开大火收汁或以淀粉水勾薄芡即可。

Ingredients:
1/2 simulate native chicken,
1 tbsp dried black fungus,
30 pieces dried lily flower, 1 bamboo shoot,
1 stalk green onion, 3～4 slices ginger

Seasonings:
3 tbsp soy sauce, 1 tbsp wine,
1 tsp sugar, 1/2 tsp salt, a little of cornstarch paste

Procedures:
1. Chop chicken to pieces, blanch for 20～30 seconds, drain and rinse to clean.
2. Soak black fungus and dried lily flower in water to soft, then trim it. Slice bamboo shoot.
3. Fry green onion sections and ginger with 2 tablespoons of oil. Add bamboo shoot slices, stir-fry for a while. Add soy sauce, wine, sugar, salt, and 2 cups of hot water. Add chicken and fungus in, stir evenly.
4. Bring to a boil over high heat, then reduce the heat to low, stew for 40 minutes. Add lily flower, continue to cook for 20 minutes until the tenderness you prefer. Season again if needed.
5. Turn to high heat to reduce the juice or you can use some cornstarch paste to thicken the sauce.

Stewed Chicken with Fresh Tomatoes

※沾了面粉再煎过的鸡会比较有香气，鸡皮的颜色也好看，但是汤汁中有了面粉和番茄糊会变得较浓稠，在烧的时候要小心，它容易粘锅底。

※Coating with flour prior to frying gives the chicken more flavor and a golden color. However, be careful that the flour and tomato paste will thicken the sauce and make it more likely to burn.

鲜茄烧鸡
Stewed Chicken with Fresh Tomatoes

材料：
鸡半只（约1.2千克）、洋葱 1/2 个、
洋菇8～10朵、番茄2个、冷冻青豆2大匙、面粉 1/2 杯

调味料：
（1）盐 1/2 茶匙、黑胡椒粉少许
（2）番茄糊1大匙、酒1大匙、淡色酱油1茶匙、水或清汤2杯、盐 1/3 茶匙、糖 1/2 茶匙

做法：
1. 鸡剁成块，放入大碗中，撒下盐和胡椒粉拌一下，放置5分钟。下锅煎之前，沾上一层面粉。
2. 番茄切刀口，放入滚水中烫一下，再泡入冷水中去皮，每个切成4小块，尽量除去番茄籽。
3. 洋菇一切为二；洋葱切粗条备用。
4. 锅中烧热2大匙油，放下鸡块煎黄外皮，盛出。放下洋葱和洋菇炒香，再放下番茄块和番茄糊一同略炒。
5. 淋下酒和酱油，加入水，煮滚后放下鸡块，加入盐和糖调味，以小火煮35～40分钟至鸡已经够烂或煮至喜爱的口感，加入青豆再煮一下。
6. 再试一下味道，适量调味。

Ingredients:
1/2 chicken (about 1.2kg.), 1/2 onion, 8～10 mushrooms, 2 tomatoes, 2 tbsp green peas, 1/2 cup flour

Seasonings:
(1) 1/2 tsp salt, pinch of pepper
(2) 1 tbsp tomato paste, 1 tbsp wine, 1 tsp light colored soy sauce, 2 cups soup stock or water, 1/3 tsp salt, 1/2 tsp sugar

Procedures:
1. Chop chicken to pieces, mix with salt and pepper, set aside for 5 minutes. Coat with a thin layer of flour just before frying it.
2. Make a cross-cut on tomato, blanch for 10 seconds, then soak in cold water quickly to peel the skin. Cut each tomato to 4 pieces. Remove seeds as possible.
3. Halve mushroom; cut onion to stripes.
4. Heat 2 tablespoons of oil to fry chicken until browned, remove. Add onion and mushrooms in, stir-fry until fragrant. Add tomatoes and tomato paste, stir evenly.
5. Add wine, soy sauce, and water, bring to a boil. Add chicken, season with salt and sugar. Simmer for 35～40 minutes to the tenderness you like. Add green peas, cook for a while.
6. Adjust the taste before turn off the heat.

Stewed Chicken with Fried Tofu in Casserole

细说全鸡

沙锅油豆腐鸡
Stewed Chicken with Fried Tofu in Casserole

材料：
仿土鸡半只（约1.2千克）、油豆腐8个、宽粉条1把、葱2根、姜2片、红辣椒1个、香菜适量

调味料：
绍兴酒1大匙、酱油4大匙、热水3杯、冰糖1茶匙、盐适量调味

做法：
1. 鸡剁成块；葱切长段；宽粉条用温水泡软；油豆腐用热水汆烫一下，捞出。
2. 起油锅，用2大匙油爆香葱段和姜片，至香气透出时加入鸡块同炒，炒至鸡块变色，淋下酒和酱油再炒一下。
3. 注入热水，再加入冰糖、红辣椒和油豆腐，一起倒入沙锅中，炖煮滚后改小火，煮1小时以上（可适量增减焖煮的时间）。
4. 至鸡肉已经快好时，加入宽粉条煮至软，如有需要可适量再加盐、糖调味。放一撮香菜装饰。

※也可以用油豆腐泡代替油豆腐来烧。
※鸡肉烹煮的时间可依照个人使用的鸡种来决定，肉鸡煮30分钟就可以了。

Ingredients:
1/2 simulate native chicken, 8 pieces fried tofu, 1 bundle mung bean noodles, 2 stalks green onion, 2 slices ginger, 1 red chili, cilantro for decoration

Seasonings:
1 tbsp shao-xing wine, 4 tbsp soy sauce, 3 cups hot water, 1 tsp rock sugar, salt to taste

Procedures:
1. Chop chicken to pieces. Cut green onion to sections. Soak mung bean noodles with warm water to soft, drain. Blanch fried tofu, remove and drain dry.
2. Heat 2 tablespoons of oil to sauté ginger and green onion until fragrant. Add chicken in, stir-fry until the color of chicken changes. Add wine and soy sauce, stir-fry for a while.
3. Add hot water, bring to a boil. Add rock sugar, red chili and fried tofu, pour all into a casserole, turn to low heat when it boils, simmer over 1 hour.
4. Cook until chicken is tender enough. Add mung bean noodles at last, cook until soft. Season with salt if needed. Put cilantro on top for decoration.

※You may use fried tofu packs instead of fried tofu.
※Cooking time differs according to the type of chicken used. For a broiler, 30 minutes will do.

Home Style Baked Chicken

细说全鸡
29

香烤手扒鸡
Home Style Baked Chicken

材料：
小土鸡、仿土鸡或肉鸡1只（约1.5千克）

腌鸡料：
葱3根、姜3片、酒2大匙、盐1大匙、黑胡椒粒1大匙、油1大匙

做法：
1. 鸡要仔细清除内脏，洗净后擦干。在鸡腿和鸡胸等肉厚的地方，用叉子叉几下，使其易入味。
2. 葱姜拍碎，和其余的腌鸡料混合，涂抹鸡身的内、外，腌6小时以上。
3. 烤箱预热至200℃，鸡放在烤盘上（腌料塞入肚内），盖一张铝箔纸，再放入烤箱中。
4. 烤30分钟之后，把鸡翻面，再烤30分钟。然后将鸡胸朝上，再烤15～20分钟（最后烤鸡胸时，可以刷上一些油，移除铝箔纸，使鸡皮上色）。
5. 烤好后取出，可用手或刀叉撕来吃。肉厚的地方如果不太咸，可以蘸胡椒盐。

※ 也可以只用鸡腿或鸡翅膀来烤，烤的时间就要缩短，大约30分钟即可。
※ 土鸡的肉比较紧，烤起来香，而普通肉鸡烤好较嫩。要选小一点的鸡，容易烤熟，且外皮不会太焦。

Ingredients:
1 small sized chicken (about 1.5kg.)

Seasonings:
3 stalks green onion, 3 slices ginger,
2 tbsp wine, 1 tbsp salt, 1 tbsp black peppercorns, 1 tbsp oil

Procedures:
1. Clean chicken thoroughly, especially the belly cavity, rinse several times and then wipe it dry. Pierce the thick part of chicken with a fork, such as legs and breast.
2. Crush green onion and ginger, mix with other seasonings. Rub all over chicken with seasonings, marinate for 6 hours. Stuff the seasonings into the belly cavity.
3. Preheat oven to 200℃, place chicken on a tray, cover with a piece of foil, put into oven.
4. Bake for 30 minutes, then turn the chicken over, bake for another 30 minutes. Place the chicken with breast side up, bake for 15~20 minutes (brush some oil over chicken breast and remove the foil to color the chicken).
5. Serve with some pepper salt as dipping sauce.

※ This dish can be made using only chicken legs and wings. The cooking time should be shortened to about 30 minutes.
※ Native chicken has tighter meat and more flavor. However, regular chicken is more tender. Smaller chickens are better for this dish because it is easier to bake through without burning the skin.

清蒸瓜子鸡汤
Steamed Chicken Soup with Pickled Cucumber

材料：
土鸡半只、酱瓜1杯、姜片3片
调味料：
酒1大匙、盐酌量、滚水或热的鸡高汤6杯
做法：
1. 鸡剁成3厘米大小的块，先用滚水汆烫30秒钟，待鸡肉已转白，即可捞出，冲洗干净。
2. 将鸡块放入大汤碗或个人用小蒸碗内（小蒸碗内只可放3~4块鸡块），加入姜片，淋下酒，再注入滚水或热高汤，放进蒸锅中，用中火蒸1.5小时。
3. 把酱瓜和半杯的酱瓜汁加入汤碗中，搅拌一下后，再继续蒸20分钟以上。
4. 关火后加入适量的盐调味即可。

※酱瓜的种类很多，应选择较无甜味的。

Ingredients:
1/2 native chicken,
1 cup Chinese pickled cucumber, 3 slices ginger
Seasonings:
1 tbsp wine, salt to taste,
6 cups hot water or hot soup stock
Procedures:
1. Cut chicken into 3cm pieces. Blanch for 30 seconds, remove and rinse.
2. Put all chicken into a large bowl, or separate them to several individual soup bowls (3 ~ 4 pieces of chicken per bowl). Add ginger, wine, and hot water (or soup stock) in, steam over medium heat for 1.5 hours.
3. Add pickled cucumber and 1/2 cup of pickled juice from the bottle to soup. Stir evenly, continue to steam for 20 minutes.
4. Turn off the heat, season with salt, serve hot.

※Choose the unsweetened pickled cucumber to make this soup.

补身精华鸡汁
Concentrate Chicken Broth

材料：
土鸡或乌骨鸡1只、姜2片

调味料：
盐少许

做法：
1. 土鸡剁成块，用滚水烫煮半分钟，捞出、冲洗一下，放入汤碗中，完全不加水，只在上面放2片姜。
2. 电饭锅中放2杯水，放入汤碗，蒸至开关跳起后，再加2杯水，再蒸至开关跳起。
3. 再加一次水继续蒸。也可用蒸锅来蒸，至少要蒸1.5小时。
4. 将原汁鸡汤倒入另一个碗中，撇除鸡油脂，可适量加盐调味。
5. 鸡块可按后面"白煮鸡的变化篇"中的食谱来做变化，或直接撒一些盐和胡椒食用。

※这种原汁的鸡汁……合体质虚弱的病人、幼儿及老年人。蒸时可以……求，加葱段或少许酒一起蒸。

Ingredients

1 native chicken
or black meat chicken,
2 slices ginger

Seasoning

a little of salt to taste

Procedures

1. Cut chicken to pieces, blanch for $1/2$ minute. Remove and rinse, put into a soup bowl. Place only ginger on top, no water at all.
2. Put 2 ~ 3 cups of water in a rice cooker, then put soup bowl in, steam until switch off.
3. Add water again, continue to steam. You need to steam it for about 1.5 hours in total (you may use steamer to steam the chicken).
4. Pour the concentrate chicken broth to a bowl, remove grease from soup. Season with a little of salt as you wish.
5. You may sprinkle some salt and pepper for the chicken meat or use it to make other kind of dishes according to the recipes I offer you in this cookbook.

※This concentrate chicken broth is very nourish. You may add some green onion and wine while steaming.

Chicken Soup with Abalone in Casserole

沙锅鲍鱼土鸡汤
Chicken Soup with Abalone in Casserole

材料：
土鸡1只、冷冻鲜鲍2粒（约250克）、干贝5~6粒、猪肉300克、大白菜600克、葱2根、姜1块

调味料：
绍兴酒2大匙、盐适量

做法：
1. 烧滚一锅开水，先把白菜整叶烫一下，取出，凉后切宽段。
2. 放下土鸡和猪肉烫煮1~2分钟，取出、冲洗干净。
3. 鲜鲍也烫一下，要刷洗去黑色黏液，再切成厚片。
4. 土鸡、猪肉、鲜鲍和干贝一起放入大沙锅中，加入葱长段和姜块，注入滚水（也可用鸡高汤，味道更好）和酒。煮滚后改小火炖煮2小时。
5. 大白菜放入沙锅中（尽量放在下面），再炖15~20分钟。加适量盐调味。

※加一块猪肉可以增加肉的香气，小蹄膀、猪脚、后腿肉或五花肉均可使用。

Ingredients:
1 native or simulate native chicken or free range chicken, 2 pieces frozen abalone (about 250g.), 5~6 dried scallops, 300g. pork, 600g. Chinese cabbage, 2 stalks green onion, 1 piece ginger

Seasonings:
2 tbsp Shao-xing wine, salt to taste

Procedures:
1. Bring a pot of water to a boil, blanch Chinese cabbage in whole pieces, remove and let it cools. Cut to wide sections.
2. Blanch chicken and pork for 1~2 minutes. Remove and rinse to clean.
3. Blanch abalone, remove and brush off the black part, cut into thick pieces.
4. Place chicken, pork, abalone, and scallops in a large casserole, add green onion sections and ginger, pour boiling water (or soup stock) and wine in, cover the lid and bring to a boil. Turn to low heat, simmer for about 2 hours.
5. Add blanched Chinese cabbage in, arrange them under the chicken as passible as you can. Cook for another 15~20 minutes. Season with salt at last.

※Add one piece of pork can enhance the fragrant, you may use pork shank, foot, rump or belly part.

细说全鸡

麻油鸡／麻油鸡面线
Chicken Soup with Sesame Oil Fragrant and Noodles

材料：
土鸡或放山鸡1只、
老姜200克、面线300克

调味料：
黑麻油1杯、米酒2杯、
冰糖1大匙、水3～4杯

做法：
1. 将鸡清理干净后切成5厘米大小的长方块；老姜不削皮，刷洗干净后切片或拍扁均可。
2. 起油锅，放下麻油和老姜，煸炒至香且黄时，放下鸡块，以大火同炒。
3. 炒至鸡皮略焦黄有香气，3～5分钟后，淋下米酒、水及冰糖，先用大火煮滚，再改用小火续煮20～25分钟，见鸡肉已熟透即熄火。
4. 面线用滚水烫煮至熟，捞出放在碗中，舀入麻油鸡和鸡汤即可食用。

※加入米酒后也可以点火烧一下，使酒精挥发掉

Ingredients:

1 native chicken
or free range chicken,
200g. ginger, 300g. thin noodles

Seasonings:

1 cup black sesame oil,
2 cups rice wine (500cc),
1 tbsp rock sugar, 3 ~ 4 cups water

Procedures:

1. Rinse chicken, then cut into 5cm pieces. Keep the skin on ginger, just brush to lean, crush or slice it.
2. In a wok, add sesame oil and ginger, stir-fry together until fragrant. Add chicken, stir-fry over high heat.
3. Fry until the color of the chicken has changed, 3 ~ 5 minutes. Add wine, water and sugar, bring to a boil over high heat. Reduce to low heat, cook for about 20 ~ 25 minutes until cooked.
4. Cook thin noodles until done, remove to a bowl, add chicken and soup, serve together with the chicken soup.

※You may flambé with rice wine to enhance the fragrant.

凤梨苦瓜鸡
Chicken Soup with Bitter Gourd & Pickled Pineapple

材料：
土鸡或半土鸡 1/2 只、苦瓜1条、腌凤梨1杯、葱1根、姜片3片

调味料：
米酒2大匙、盐适量、胡椒粉少许

做法：
1. 鸡剁成块，用滚水氽烫一下，捞出洗净。
2. 苦瓜剖开、去籽，切成厚块状。
3. 汤锅中煮滚6～7杯水，放下鸡块和葱、姜、酒，煮滚后改小火，炖煮约30分钟。
4. 将腌凤梨连汁一起加入汤中，再以小火煮约20分钟。加入苦瓜，再煮至喜爱的烂度。
5. 尝过味道，看是否需要加盐调味，撒下胡椒粉即可。

※腌凤梨的咸度不同，因此最后要尝一下味道再加适量的盐调味

Ingredients:
1/2 native or simulate native chicken,
1 bitter gourd, 1 cup pickled pineapple,
1 stalk green onion, 3 slices ginger

Seasonings:
2 tbsp wine, salt to taste, pinch of pepper

Procedures:
1. Cut chicken to pieces, blanch for a while, remove and rinse.
2. Split bitter gourd, remove seeds, then cut into thick pieces.
3. Bring 6~7 cups of water to a boil, put chicken, green onion, ginger, and wine in. Bring to a boil, turn to low heat, simmer for about 30 minutes.
4. Add pickled pineapple into soup, cook for 20 minutes over low heat. Add bitter gourd in, cook to the tenderness you like.
5. Season with some salt if needed, sprinkle some pepper. Serve.

※pickled pineapple is very salty, taste the soup before adding salt

细说全鸡

韩式参鸡汤
Ginseng Chicken Soup, Korean Style

材料：
小土鸡1只、新鲜人参1枝、糯米4大匙、大蒜4粒、红枣4~5粒、松子1大匙

调味料：
盐适量、胡椒粉少许

做法
1. 鸡整只洗净，挖除内部血块及内脏。
2. 糯米洗净、泡水20分钟；人参、红枣冲洗干净，人参太大时可以切开成小条。
3. 将糯米和人参条、2粒红枣及2粒大蒜塞入鸡的腹中，用竹签封住开口。
4. 炖锅中放入6杯水，煮滚后放入人参鸡、剩下的人参条、红枣和大蒜粒，大火煮滚后改小火煮约1.5小时。
5. 撒下松子，并酌量加盐和胡椒粉调味，附胡椒盐上桌蘸食鸡肉。

Ingredients:
1 small native chicken, 1 stalk fresh ginseng, 4 tbsp glutinous rice, 4 cloves garlic, 4~5 red dates, 1 tbsp pine nuts

Seasonings:
salt to taste, pinch of pepper

Procedures
1. Rinse chicken, remove internal organs.
2. Rinse glutinous rice, soak for 20 minutes; rinse ginseng and red dates; you may split ginseng if it is too large.
3. Stuff rice, 2 slices of ginseng, 2 red dates, and 2 cloves garlic into the cavity of chicken belly. Seal with toothpicks.
4. Bring 6 cups of water to a boil, put chicken and the remaining ginseng, red dates and garlic in, simmer for 1 $1/2$ hours.
5. Add pine nuts in, season with salt and pepper, serve with some pepper salt to dip the chicken meat.

赤小豆山药煲鸡汤
Chicken Soup with Red Beans and Yam

材料：
乌骨鸡1/2只、山药450克、干鱿鱼或干章鱼50克（1/3条）、
蜜枣1颗、红豆1/4杯、姜1块

调味料：
酒2大匙、盐适量

做法：
1. 乌骨鸡剁成块，用滚水汆烫1分钟，捞出、洗净。
2. 蜜枣用温水洗一下；干鱿鱼剪成小块，用水冲一下；红豆冲洗一下。
3. 将上面4种材料和姜放入汤锅中，加入滚水7～8杯和酒，煮滚后改小火煲煮1.5小时以上。
4. 山药切块，放入汤中，再煮8～10分钟，加适量的盐调味。

Ingredients:
1/2 black-meat chicken,
450 Chinese yam,
50g. dried squid or octopus
(about 1/3 piece),
1 brown date,
1/4 cup small red beans, 1 piece ginger

Seasoning:
2 tbsp wine, salt to taste

Procedures:
1. Cut chicken to pieces, blanch for 1 minute. Remove and rinse to clean.
2. Rinse date with warm water; cut squid to small pieces with kitchen scissors, rinse; rinse red beans.
3. Put the four ingredients which mentioned above into a soup pot with ginger, add 7～8 cups of hot water and wine in, bring to a boil. Reduce heat to low, simmer for about 1.5 hours.
4. Cut Chinese yam to pieces, cook with soup for 8～10 minutes, season with salt.

Boiled Chicken

※可以把姜、葱末放小碗中，淋下热油，再调入盐，做成葱油蘸汁。

※Pour heated oil into chopped green onion and ginger to make dipping sauce().

细说全鸡

白煮鸡
Boiled Chicken

材料：
土鸡、半土鸡或小放山鸡1只（2~2.4千克）、姜1块、酒2大匙

蘸料：
（1）葱细末1大匙、姜末1茶匙、热油2大匙
（2）酱油膏1大匙

做法：
1. 鸡内部的血块要清洗干净。
2. 挑一个煮鸡的汤锅，锅的口径不要太宽，以免水太多，使鸡的鲜味流失，锅中的水量要刚好淹盖过鸡约2厘米，先把水煮滚，加入酒和姜片。
3. 抓住鸡脖子，把鸡身浸入水中，5秒钟后提出，待水再煮开，再烫一次，最好鸡的腹部也灌一次热水。放入鸡，等水再煮滚后，改成小火煮18~20分钟。
4. 关火后把鸡泡在鸡汤中约30分钟，捞出，用竹签试一下肉厚的腿部，如没有血水流出即是熟了，取出鸡，放在盘子上。盖上一条湿毛巾散热，待凉后抹上鸡汤上的鸡油或麻油以保持鸡的香气。
5. 如要使鸡皮发脆，可以把鸡取出后放在盆中，加入冰块和冰水盖过鸡（见第16页的小图），泡至鸡已凉透（约30分钟），取出、抹油。
6. 鸡肉剁块，淋2~3大匙的鸡汁，附上葱姜油和酱油膏蘸食鸡肉。

Ingredients:
1 native chicken or free range chicken or simulate native chicken (2~2.4kg.),
1 piece ginger, 2 tbsp wine

Dipping sauce:
(1) 1 tbsp finely chopped green onion, 1 tsp chopped ginger, 2 tbsp heated oil
(2) 1 tbsp soy sauce paste

Procedures:
1. Clean the cavity of chicken belly thoroughly.
2. Choose a pot which can fit the chicken, do not use an oversized pot, or you will loose the taste of chicken. Bring the water to a boil (the water should be 2cm higher then chicken), add wine and ginger in.
3. Hold the neck of chicken, sunk chicken into water for 5 seconds, remove. Bring water to a boil, blanch again. Pour some hot water into the belly. Put chicken in after water boils, cook until water boils again, turn to low heat, simmer for 18~20 minutes.
4. Turn off the heat, cover for 30 minutes. Remove chicken. Try to pierce the thickest part of the chicken by a sharp stick, it is cooked if the juice is cleared. Remove to a plate, cover the chicken with a wet towel to let it cools. Rub some chicken grease or sesame oil over chicken.
5. If you wish to have a crispy and chewy skin, you may put the cooked chicken in a large pot, add ice and ice water to cool down chicken quickly (as the picture on P.16). Soak for about 30 minutes. Remove and then brush some oil.
6. Cut chicken to pieces, arrange on a plate, pour 2~3 tablespoons of broth over chicken, serve with dipping sauce (1) and (2).

五香盐水鸡
Salty Chicken

材料：
白煮鸡1只
盐水卤：
葱2根、姜1块、红葱头2～3粒、大蒜2粒、五香包1包、
绍兴酒2大匙、盐4大匙、水（或煮鸡的汤）4杯
做法：
1. 锅中加热油2大匙，爆香大蒜、姜片、红葱头和葱段，待香气透出时，淋下酒炒煮一下，放入五香包、水和盐大火煮滚，改小火煮15～20分钟，做成五香盐水卤汤（2～2 1/2 杯），放凉备用。
2. 白煮鸡自锅中取出后放入冰块中（请参考前页白煮鸡），倒下盐水卤，再加入冰水直到淹没鸡，放入冰箱中浸泡4～5小时至入味。
3. 吃时剁成块装盘，淋上盐水卤。盐水鸡的胸肉也可以撕成条或剁成块来做凉拌菜。

Ingredients:
1 boiled chicken
Salty Spicy Broth:
2 stalks green onion,
1 piece ginger, 2 ~ 3 cloves shallot,
2 cloves garlic, 1 five-spicy pack,
2 tbsp Shaoxing wine,
4 tbsp salt, 4 cups water or chicken broth
Procedures:
1. Heat 2 tablespoons of oil to sauté garlic, ginger, shallot, and green onion until fragrant. Add wine, boil for a while. Add five-spicy pack, water and salt, bring to a boil. Turn to low heat, cook for 15 ~ 20 minutes to make the salty spicy broth (about 2 ~ 2 1/2 cups). Let it cools.
2. Remove chicken from heat, put into ice cubes (as the recipe shows on P.35), add the salty spicy broth, cover the whole chicken with ice water. Remove to fridge, soak for 4 ~ 5 hours.
3. Cut to pieces before serving, pour some broth over chicken. Also you may make other salad dishes with the remaining chicken breast.

醉鸡
Wined Chicken

材料：
白煮鸡腿2只、白煮鸡翅2只
醉鸡浸汁：
绍兴酒1杯、冷鸡汤2杯、
鱼露或虾油 2/3 杯、花椒粒 1/4 茶匙
做法：
1. 按照第40页的方法，把白煮鸡煮好，放凉，切下要做醉鸡的腿和鸡翅部分（要整块）。
2. 醉鸡浸汁在大碗内混合、调好味道，放入鸡块，鸡块要完全浸入酒汁中，移入冰箱内，浸泡半天以上（最好泡24小时）。
3. 吃时剁成小块，或者把大骨去除后再切块装盘。

※浸的酒汁要盖住鸡块，所以用一个盘子压在鸡块上。

Ingredients:
2 boiled chicken legs,
2 boiled chicken wings
Seasonings:
1 cup Shaoxing wine,
2 cups cold chicken stock,
2/3 cup fish sauce or shrimp sauce,
1/4 tsp brown peppercorns
Procedures:
1. Cook chicken according to P.35. Cut chicken legs and wings off (keep in whole shape) when cools.
2. Mix seasonings in a large bowl to make wine broth. Soak chicken in, the broth must cover all the chicken pieces. Remove to fridge for at least 12 hours (it will taste better if you soak it for 24 hours).
3. Cut to pieces before serving, or you may remove the bones before cutting it.

※You may cover the chicken is soaked in wine broth completely.

口水鸡
Mouth Watering Chicken Salad, Sichuan Style

材料：
白煮鸡1/2只、干辣椒1大匙、姜汁1/2茶匙、蒜茸1茶匙、葱花1大匙、熟花生米2大匙、芝麻1/2大匙、花椒粒1大匙

调味料：
酱油2大匙、醋2茶匙、盐1/4茶匙、糖2茶匙、麻油1茶匙、煮鸡汤2大匙

做法：
1. 鸡剔下大骨，剁成小块，垫在盘底；肉切直条，铺放在上面。
2. 干辣椒和2大匙油一起放锅中，开火煸炒，以小火炒至香气透出，盛出，剁碎。
3. 花椒粒用1 1/2大匙油小火炒香，捞弃花椒粒，油留用。
4. 花生米略压碎；芝麻炒过，放凉。
5. 干辣椒碎、姜汁、蒜茸、葱花放碗中，加入花椒油、辣椒油和所有调味料调匀，淋在鸡肉上，撒上花生碎和芝麻，上桌后拌匀。

※盘底可垫一些生菜丝。

Ingredients:
1/2 cooked chicken, 1 tbsp dried red chili, 1/2 tsp ginger juice, 1 tsp mashed garlic, 1 tbsp chopped green onion, 2 tbsp fried or roasted peanuts, 1/2 tbsp sesame seeds, 1 tbsp brown peppercorns

Seasonings:
2 tbsp soy sauce, 2 tsp vinegar, 1/4 tsp salt, 2 tsp sugar, 1 tsp sesame oil, 2 tbsp chicken soup from cooked chicken

Procedures:
1. Remove the big bones from chicken, chop to small pieces, place on the plate. Cut the meat, place on top of bones.
2. Put dried chili in a wok with 2 tablespoons of oil, turn on the heat, fry over low heat until fragrant. Remove and chop finely.
3. Fry brown peppercorns with 1 1/2 tablespoons of oil, discard peppercorns, keep the oil.
4. Crush peanuts a little, stir-fry sesame seeds over low heat, remove, let cools.
5. Mix dried chili, ginger juice, garlic, green onion, chili oil and brown pepper oil with seasonings in a bowl. Pour over chicken, sprinkle peanuts and sesame seeds over. Serve.

※You may place some shredded lettuce on the bottom.

蜇皮手撕鸡
Chicken and Jellyfish Salad

材料：
白煮鸡腿1只或鸡胸1个、海蜇皮150克、黄瓜1条、葱丝 1/2 杯、嫩姜丝2大匙、香菜段 1/3 杯、白芝麻1大匙

调味料：
酱油2大匙、麻油1大匙、
醋 1/2 大匙、盐 1/4 茶匙、糖1茶匙

做法：
1. 白煮鸡的鸡肉用手撕成粗丝。
2. 海蜇皮切丝，用水冲洗几次后，用冷水泡1～2小时。
3. 在锅中放5杯水，煮至8分热，放入海蜇皮烫3～5秒钟，捞出再泡入冷水中，至海蜇丝涨大。用冷开水冲洗过，沥干后再以纸巾擦干。
4. 黄瓜切丝，放入大碗中，加入鸡丝、海蜇、葱丝、嫩姜丝和香菜段，淋下调匀的调味料，拌匀后装盘，撒下炒过的白芝麻即可。

Ingredients：
1 cooked chicken leg or breast,
150g. dried jellyfish,
1 cucumber,
1/2 cup green onion shreds,
2 tbsp ginger shreds,
1/3 cup cilantro sections,
1 tbsp fried sesame seeds

Seasonings：
2 tbsp soy sauce, 1 tbsp sesame oil,
1/2 tbsp vinegar, 1/4 tsp salt, 1 tsp sugar

Procedures：
1. Tear chicken meat to strips.
2. Shred jellyfish, rinse for several times. Soak in water for 1~2 hours.
3. Bring 5 cups of water to 80℃, blanch jellyfish for 3~5 seconds. Remove and soak in cold water until jelly fish expand. Rinse with drinking water, drain and wipe dry.
4. Shred cucumber, place in a large bowl with chicken shreds, jellyfish, green onion, ginger and cilantro. Pour mixed seasonings in, mix evenly and then remove to a plate, sprinkle sesame seeds over the top.

白煮鸡的变化

东安鸡
Stir-fried Chicken, Dongan Style

材料：
白煮鸡 1/2 只、葱丝 1/2 杯、
姜丝 1/3 杯、红辣椒丝 1/4 杯、花椒粒 1/2 大匙

调味料：
（1）酒1大匙、酱油2大匙、糖1茶匙、盐 1/4 茶匙、煮鸡汤 1/2 杯
（2）醋1大匙、淀粉水 1/2 大匙、麻油数滴

做法：
1. 白煮鸡如用鸡胸部分，则先去除大骨，再剁成块；如果用鸡腿，就直接剁成块。
2. 炒锅内放2大匙油和花椒粒，开火加温，炒到花椒变色且有香气后捞弃。
3. 接着加入一半量的葱丝、姜丝和红椒丝略爆炒，随后放下鸡块，依序加入调味料（1），煮3分钟。
4. 再沿锅边淋下醋增香，用淀粉水勾芡，最后淋下麻油，撒下剩下的葱丝，炒匀即可装盘。

Ingredients：
1/2 cooked chicken,
1/2 cup shredded green onion,
1/3 cup shredded ginger,
1/4 cup shredded red chili,
1/2 tbsp brown peppercorns

Seasonings：
(1) 1 tbsp wine, 2 tbsp soy sauce,
  1 tsp sugar, 1/4 tsp salt,
  1/2 cup stock from cooked chicken
(2) 1 tbsp vinegar, 1/2 tbsp cornstarch paste,
  a few drops of sesame oil

Procedures：
1. You may choose any part of cooked chicken, remove large bone from breast, then cut to pieces. If you use the chicken leg, just chop to pieces with bones.
2. Heat 2 tablespoons of oil with brown peppercorns, remove peppercorns when fragrant and the color turns to dark brown.
3. Add half amount of green onion shreds, ginger and red chili in, fry until the aroma come out. Add chicken and seasonings (1), stir-fry over high heat, then simmer for 3 minutes.
4. Splash vinegar around the edge of wok, thicken with cornstarch paste. Drizzle sesame oil in, then add remaining green onion. Mix evenly and remove to a plate.

白煮鸡的变化

Flavored Chicken Salad

白煮鸡的变化

怪味鸡
Flavored Chicken Salad

材料：
白煮鸡 1/2 只、姜末2茶匙、
蒜泥2茶匙、香菜2根、生菜1棵
调味料：
酱油膏3大匙、糖1茶匙、醋 1/2 大匙、
芝麻酱2茶匙、煮鸡汤2大匙、花椒粉1茶匙、红油2茶匙
做法：
1. 白煮鸡连骨剁成长条块。
2. 生菜洗净，铺放在盘底，上面放上鸡块。
3. 芝麻酱先调稀，再加上姜末、蒜泥和其余的调味料调匀，淋在鸡块上，再撒上香菜末。

※这是一道四川名菜，集酸、咸、麻、甜、香、辣各种味道，十分特殊。

Ingredients：
1/2 cooked chicken, 2 tsp mashed garlic, 2 tsp chopped ginger, 2 stalks cilantro, 1 stalk lettuce
Seasonings：
3 tbsp soy sauce paste,
1 tsp sugar, 1/2 tbsp vinegar,
2 tsp sesame seeds paste,
2 tbsp chicken soup or water,
1 tsp brown pepper powder,
2 tsp red chili oil
Procedures：
1. Cut chicken to pieces with bones.
2. Rinse lettuce, then shred it, place on a plate, then arrange chicken on top.
3. Dissolove sesame seeds paste with chicken soup, then mix with ginger, garlic and other seasonings. Mix evenly, pour over chicken. Sprinkle cilantro sections over the top.

※This is a famous dish in Sichuan cuisine. This dish is unique because it tastes sour, salty, sweet, spicy, and fragrant at one dish.

棒棒鸡
Bon Bon Chicken

材料：
白煮鸡鸡胸1个、新鲜粉皮2张、小黄瓜1条、姜汁1茶匙、蒜泥 1/2 大匙、细葱花 1/2 大匙

综合调味料：
芝麻酱1 1/2 大匙、酱油2大匙、水1大匙、镇江醋1大匙、麻油1大匙、辣椒油1茶匙、糖2茶匙、盐 1/4 茶匙、花椒粉 1/2 茶匙

做法：
1. 黄瓜切成片，撒少许盐腌约10分钟。冲洗一下，挤干水分，铺放在盘中。
2. 粉皮切成宽条，用冷开水漂洗一下，捞出沥干，放在盘内黄瓜的片上。
3. 在白煮鸡的胸骨两边各切一道刀口，取下鸡胸肉，太厚的地方再片切一刀成为两片，用刀面拍一下，使鸡肉松弛一点，再切成条，排列在粉皮上。
4. 小碗中，先放芝麻酱，再慢慢加入酱油和水，调开成稀稠糊状。再陆续加其他的调味料和姜汁、蒜泥及葱花，待上桌前浇到鸡肉上，吃时略加拌匀便可。

※ 棒棒鸡垫底的材料可以加以变化，例如莴笋、西芹、百页豆腐、素鸡等均很适合。
※ 可以用干粉皮代替新鲜粉皮，先用水将粉皮泡软，再放入滚水中煮至透明，捞出后冲凉，沥干水分即可使用。

Ingredients:
1 cooked chicken breast, 2 fresh mung bean sheets, 1 cucumber, 1 tsp ginger juice, 1/2 tbsp mashed garlic, 1/2 tbsp finely chopped green onion

Seasonings:
1 1/2 tbsp sesame seed paste, 2 tbsp soy sauce, 1 tbsp water, 1 tbsp brown vinegar, 1 tbsp esame oil, 1 tsp red chili oil, 2 tsp sugar, 1/4 tsp salt, 1/2 tsp brown pepper powder

Procedure:
1. Slice cucumber, mix with a little of salt for about 10 minutes. Rinse and squeeze out the salty juice, lay on a plate.
2. Cut bean sheets into 1/2 " wide and rinse with cold water. Drain and arrange on top of cucumber.
3. Remove chicken bones, cut the meat into stripes, place on top of bean sheets.
4. Mix sesame seeds paste with soy sauce and water little by little, then mix with all other seasonings. Add ginger, garlic and green onion, mix well, serve with chicken. Pour the sauce over chicken and mix carefully on table before eating.

※ You may use green bamboo shoot, celery, firmed tofu, or vegetarian chicken instead of cucumber and mung bean sheet.
※ Dried mung bean sheets can be used instead of the fresh ones. Soak and soften the dried sheets first, and then place the sheets in boiling water until they turn transparent. Remove and rinse until cool. Drain off the excess water before use.

三蔬拌鸡丝
Chicken Salad with Vegetables

材料：
白煮鸡 1/4 只、甜豌豆片100克、金针菇 1/2 包、胡萝卜丝 1/2 杯、大蒜酥2茶匙

调味料：
虾油1茶匙、盐 1/4 茶匙、麻油1茶匙、沙茶酱1茶匙、煮鸡汤1大匙

做法：
1. 白煮鸡切成丝或撕成丝。
2. 豌豆片用滚水（水中加少许盐）氽烫一下，捞出、泡入冰水中冰凉，再取出、擦干、切成细丝。
3. 金针菇切除根部，再切成两段，也入滚水中氽烫一下，冲凉、挤干水分。
4. 鸡丝和三种蔬菜在碗中和调味料拌匀，装入盘中，撒下大蒜酥。

※大蒜酥可买现成的，再剁碎一点，或把大蒜剁碎之后用油炸香，连蒜油一起用。

Ingredients:
1/4 cooked chicken,
100g. snow pea pots,
1/2 pack needle mushrooms,
1/2 cup carrot shreds,
2 tsp fried chopped garlic

Seasonings:
1 tsp shrimp sauce, 1/4 tsp salt,
1 tsp sesame oil, 1 tsp sha-cha sauce,
1 tbsp chicken soup or drinking water

Procedures:
1. Tear or shred the chicken.
2. Blanch snow pea pots (add a little of salt in water), remove and soak in iced water. Remove, drain dry and then cut to strings.
3. Trim needle mushrooms, cut into two parts. Blanch and rinse, squeeze dry.
4. Mix chicken with 3 kinds of vegetable and seasonings, remove to a plate. Sprinkle fried garlic over the top.

※You may use store-bought fried garlic and chop it more finely. You may also deep-fry chopped fresh garlic and use the oil instead of sesame oil.

鸡丝凉面
Cold Noodles with Chicken

材料：
白煮鸡胸半个、绿豆芽100克、凉面300克或细面条250克、黄瓜丝1杯

调味料：
芝麻酱2大匙、酱油2大匙、冷开水2大匙、醋1/2大匙、糖1/2茶匙、蒜泥2茶匙、姜汁1/2茶匙、辣椒油1/2大匙、麻油1/2大匙、花椒粉1/4茶匙、细葱花1大匙

做法：
1. 鸡胸去骨后切成细丝。豆芽烫熟后捞出，冲凉后挤干。
2. 把凉面、豆芽、黄瓜排入盘中，再放上鸡丝。
3. 芝麻酱用酱油及冷开水慢慢地分次加入、调稀，再加入其他的调味料调匀。
4. 麻酱汁淋在凉面上，可以撒上切碎的花生米或白芝麻增加香气。

※如用生面条，要先把面条煮熟，捞出后放在大盘子里，拌麻油和油，快速吹凉。

Ingredients：
1/2 cooked chicken breast, 100g. mung bean sprouts, 300g. uncooked noodles or 250g. cooked noodles, 1 cup cucumber shreds

Seasonings：
2 tbsp sesame seeds paste, 2 tbsp soy sauce, 2 tbsp water, 1/2 tbsp vinegar, 1/2 tsp sugar, 2 tsp mashed garlic, 1/2 tsp ginger juice, 1/2 tbsp red chili oil, 1/2 tbsp sesame oil, 1/4 tsp brown pepper powder, 1 tbsp finely chopped green onion

Procedures：
1. Remove bones from chicken breast, then shred it. Blanch bean sprouts, remove and rinse to cold, squeeze dry.
2. Place cold noodles, bean sprouts, cucumber on a serving plate, put chicken on top.
3. Mix sesame seeds paste with soy sauce and water little by little until mixed. Add other seasonings in, mix evenly.
4. Pour the mixed seasonings over noodles, you may sprinkle some chopped roasted peanuts or sesame seeds on top to enhance the fragrant.

※If using fresh noodles, cook the noodles in boiling water, mix in sesame oil and vegetable oil, and let it cool quickly.

白煮鸡的变化

泰式凉拌鸡丝
Chicken Salad, Thai Style

材料：
熟鸡胸肉1片、洋葱丝 1/3 杯、
绿豆芽1把、黄瓜丝1杯、
小番茄10颗、红葱头1～2粒、九层塔叶数片

调味料：
柠檬汁3大匙、白糖2大匙、鱼露 1/2 大匙、泰式辣椒酱2茶匙

做法：
1. 鸡胸肉切成丝；绿豆芽快速烫一下，捞出、冲凉。
2. 小番茄切半或一切为四；红葱头剁碎。
3. 将红葱头、小番茄和调味料一起混合，搅拌均匀。
4. 洋葱丝、绿豆芽、黄瓜丝和一半的调味料拌匀，放在盘子里，上面堆放鸡丝，再淋下剩余的调味料，放上九层塔。

Ingredients:
1 piece cooked chicken breast,
1/3 cup shredded onion,
1 hand full mung bean sprouts,
1 cup shredded cucumber,
10 cherry tomatoes,
1 ~ 2 shallots, a few pieces of basil leaf

Seasonings:
3 tbsp lemon juice, 2 tbsp sugar,
1/2 tbsp fish sauce, 2 tsp Thai style chili sauce

Procedures:
1. Shred chicken meat; blanch mung bean sprouts quickly, remove and rinse to cool.
2. Halve or quarter the tomato; chop shallot.
3. Combine shallot, tomatoes and seasonings.
4. Mix onion, bean sprouts and cucumber with half amount of seasonings, remove to a serving plate. Put chicken shreds on top, pour remaining seasonings over chicken. Place basil on top, serve.

白煮鸡的变化

马铃薯鸡肉沙拉
Chicken & Potato Salad

材料：
马铃薯2个（约400克）、鸡蛋3个、白煮鸡胸肉1个、
西芹2棵、美奶滋4～5大匙

调味料：
盐 1/2 茶匙、胡椒粉少许

做法：
1. 马铃薯和鸡蛋洗净，放入锅中，加水煮熟，煮约12分钟时取出鸡蛋，蛋白切小丁、蛋黄捏碎。
2. 马铃薯再煮至没有硬心，取出、剥皮、切成小块。
3. 鸡胸肉切厚片；西芹削去老筋、用滚水烫一下，冲凉后切片。
4. 所有的材料放在大碗中，加入调味料和美奶滋（沙拉酱的一种）拌匀，可留一些蛋黄做装饰。

Ingredients：
2 potatoes (about 400g.),
3 eggs, 1 piece cooked chicken breast,
2 pieces celery, 4～5 tbsp mayonnaise

Seasonings：
1/2 tsp salt, pepper to taste

Procedures：
1. Rinse potatoes and eggs. Cook until done, remove eggs after cooked for 12 minutes. Dice egg white and mash egg yolk.
2. Remove potatoes when soft. Peel and cut to small pieces.
3. Cut chicken meat; trim celery, then blanch for 10 seconds, rinse to cool and then slice it.
4. Place all ingredients in a large bowl, mix with seasonings and mayonnaise. Keep some egg yolk for decoration

芥末鸡肉沙拉
Chicken Salad with Mustarch Dressing

材料：
白煮鸡胸1个、洋葱碎 1/4 杯、西芹2棵、绿花椰菜1棵、核桃 1/3 杯（略切碎）、
泡面 1/2 片、奶油2大匙

调味料：
黄色芥末酱 1/2 大匙、美奶滋3大匙、橄榄油1大匙

做法：
1. 白煮鸡胸肉去骨、去皮，切成小块。
2. 西芹摘好后切片；绿花椰菜摘好，放入滚水中余烫，烫半分钟后放入西芹，快速烫一下即捞出一起冲凉。
3. 锅中放奶油、核桃和泡面，小火慢慢炒2分钟，盛出放凉。
4. 调味料先调匀，放下鸡肉、洋葱、西芹和绿花椰菜一起再调拌均匀，放1小时使味道融合。
5. 吃时加入核桃和面条拌匀。

Ingredients:
1 cooked chicken breast,
1/4 cup chopped onion,
2 sticks celery, 1 broccoli,
1/3 cup walnut (slightly chopped),
1/2 piece instant noodle, 2 tbsp butter

Seasonings:
1/2 tbsp yellow French mustard,
3 tbsp mayonnaise, 1 tbsp olive oil

Procedures:
1. Remove skin and bones from cooked chicken, then cut into small pieces.
2. Trim and then slice celery; trim broccoli. Blanch broccoli for 30 seconds, add celery in, blanch shortly, remove and rinse to cool.
3. Combine butter, walnut, and instant noodle in a pan, sauté over low heat for 2 minutes, remove and let it cool.
4. Mix seasonings in a bowl, then add chicken, broccoli, celery and onion in, mix and set aside for one hour to combine the taste.
5. Add walnut and noodles just before serving.

白煮鸡的变化

炖鸡煨面
Chicken Soup Noodles

材料：
白煮鸡 1/4 只、剩余白煮鸡骨架、
葱1根、小油菜200克、面条300克
调味料：
酒1大匙、盐适量
做法：
1. 白煮鸡可以只取鸡肉，把鸡肉撕成粗条块，或者连骨斩剁成小块。
2. 锅中用1大匙油把葱段煎香，淋下酒和6杯水（或煮鸡时剩下的清汤），再放下鸡骨架。
3. 大火煮滚后改成小火熬煮20～30分钟，捞除鸡骨。
4. 面条在滚水中煮一滚即捞出。
5. 小油菜切成段，和面条及鸡块一起放入汤中，煮至面条已软即可（如只用鸡丝，则不用太早放入，最后1分钟再放入汤中即可）。
6. 加适量的盐调味。

Ingredients:
1/4 cooked chicken,
remaining chicken bones
from cooked chicken,
1 stalk green onion,
200g green cabbage,
300g. noodles
Seasonings:
1 tbsp wine, salt to taste
Procedures:
1. You may tear the cooked chicken meat to strips, or cut it to pieces with the bones.
2. Stir-fry green onion sections with 1 tablespoon of oil until fragrant, splash wine, then pour water (or the remaining soup from cooked chicken) in. Add chicken bones.
3. Bring to a boil over high heat, then reduce to low heat, cook for 20～30 minutes.
4. Cook noodles in boiling water until water boils again. Remove noodles.
5. Cut green cabbage to sections, put into soup with noodles and chicken, cook until noodle is soft enough. (cook only for 1 minute if you use the chicken strips).
6. Season with salt, serve hot.

白煮鸡的变化

细说鸡胸
All about Chicken Breasts

鸡胸的分类与切法：
　　鸡胸（图1）仔细分起来可以分成鸡柳条肉、鸡胸肉和鸡胸骨架三个部分（图2），因为鸡胸肉属于白肉，脂肪的含量很低，因此以健康的角度来看，它是最健康的肉类，许多鸡肉加工品都是用它来做的。

Cuts of chicken breasts:
　　Technically the breast part of a chicken (pic 1) includes three parts: tender, breast, and bones (pic 2). Chicken breasts are lean and very low in fat; therefore they are considered the healthier meat and are commonly used in processed food products.

图1

图2

1. 柳条肉：
　　柳条肉也常称为鸡的里脊肉，是位于鸡胸肉里面的长条型的肉；一边有一条（图3），肉质很软嫩，但是切丝来炒时很容易断裂、散开，不好看，也比较没有口感。但是中式菜中的"鸡茸"，用它来做是最适合的，把它压成鸡泥（图4），再加蛋白搅匀，做菜或做羹汤、汤面都非常滑嫩可口。从前没有单独出售时，常把它冷冻保存起来，累积有4～5条的量时再拿来用，现在在超市中都有一整盒鸡柳条肉出售，价钱十分便宜。

1. Chicken tender:
The tender loins of chicken. There is one chicken tender on each side (pic 3). Chicken tender tends to fall apart when it is cut into strings. However it is most suitable for making "minced chicken" dishes. You can achieve very smooth and tender texture by mashing the chicken tender and then combining it with egg white, you may use it to make a dish or soup or potage or just add it to the soup noodle.

Chicken tenders were not packaged for sale by themselves earlier, and I used to collect them and keep them in the freezer until there were enough to make a dish. Nowadays you may find chicken tender packets in supermarkets that are cheap and convenient.

图3

图4

2. 鸡胸肉：

　　一个鸡胸可以取下两片鸡胸肉，烹调之前通常是把它切丝、切条、切片、切块或整片做鸡排。在切之前都要先修整一下，要去掉鸡皮、修掉软骨或筋膜（图5），现在在超市中买的鸡胸肉都是已经处理好的。

　　要切丝的胸肉，要先切成大薄片（图6），再切成丝或条（图7）。因为鸡肉比较嫩，因此可以顺丝切，顺着纹路切出来的鸡丝比较漂亮。

　　可以把处理好的鸡胸肉放在冰箱中冷冻，待鸡肉冻硬了再拿出来切，就很容易下刀了。现在超市中已经可以买到处理好的鸡片、鸡丝或绞鸡肉，比较方便。（图8、图9、图10）

2. Chicken breast:

There are two pieces of chicken breast meat in one chicken. Chicken breasts may be cut into strings, cubes, or slices prior to cooking, or you may keep it in whole piece to make chicken steak. The skin, cartilage, or vein should be trimmed and removed during the preparation process (pic 5). Or you may choose to purchase the skinless and boneless chicken breasts from supermarkets. Picture 6 and 7 illustrates how to cut chicken breast into strings. The breast is first sliced into large and thin slices, and then cut into strings with the grain. In this way the chicken strings will have better appearance when cooked. You may freeze the breast prior to cutting so that it is not as slippery. Chicken slices, strings, and ground chicken are also available in many supermarkets now (pic 8, pic 9, pic 10).

图5　图6　图7
图8　图9　图10

3. 鸡胸骨架：

鸡胸肉如果没有鸡架子撑着一起煮，胸肉会缩，因此做某些菜式时我们要连着鸡胸骨一起去煮，鸡胸肉的丝条才会长而好看。鸡架子常用来熬煮高汤，但是如果要鸡汤有香气，最好是连胸肉一起煮，两三个鸡架子中有一个是带着肉的，才会使鸡高汤更香。

3. chest bones (also called rib cages):

Chicken breast will shrink significantly while cooking if it is removed from the chest bones. Therefore for the presentation, chicken breast is better to cooked with the bones. Chest bones are commonly used to make chicken stock. Two or three chest bones are generally required to make a batch of chicken stock, and one chicken breast should remain on one of the chest bones in order to achieve good aroma and flavors.

鸡胸肉的腌泡方法：
　　基本上鸡胸肉要保持白色才漂亮，因此要炒或是烫之前，多半是用盐和水先抓拌，使鸡肉吸水膨胀了，再加入淀粉拌匀（图11）。如果要加蛋白，水的分量就要减少些，也可依个人喜好，适量加些酒或白胡椒、麻油拌匀来腌。如果是腌拌鸡丝时，就要轻轻地小心抓拌，不要把鸡丝弄断了。
　　腌的时候最好放进冰箱中冷藏保鲜。如果加了蛋白一起腌，最好能放置1小时以上；没有放蛋白时，最好也要放置30分钟以上，以使淀粉稳定地包裹在鸡肉外层，以便在过油时形成一层薄膜，增加肉质滑嫩的口感，同时能包裹住鸡肉，使鸡汁不外溢，保持鸡肉的鲜味。
　　用整片鸡胸烹调时，可以在胸肉上划切一些刀口再腌，以使胸肉容易入味（图12）。

How to marinate：
Chicken breast is commonly marinated in order to maintain its natural color. Salt and water are first added to the chicken meat first. After the water has been absorbed by the meat, egg white and cornstarch are added and combined (pic11). If you prefer, you may choose to marinate with wine, white pepper, or sesame oil. Be careful not to tear the chicken strings during marinating and mixing.

The marinating process should take place in a refrigerator. If you are marinating with egg white, the meat should be marinated for more than an hour; without egg white, it should be longer than 30 minutes. During marinating, cornstarch forms a thin coating outside the chicken meat, which not only creates a smooth and silky texture after the meat is cook but also keeps the chicken juice inside.

When cooking with a whole piece of chicken breast, you may cut small openings on the meat to help the marinating process(pic12).

图11

图12

烹调重点：
过油：
　　要炒的鸡肉，传统上都要先经过过油的程序，以较多量的油炒到八九分熟，再和配料、辛香料与调味料去炒。经过热油先去过油、快炒之后，可以保持鸡肉的多汁与滑嫩，同时肉也更有香气。

　　过油时所用油的量要比鸡肉多，把油先烧到6～8分热，鸡丝6～7分热；鸡片、鸡块约为7～8分热，放下鸡肉，用筷子快速搅散鸡丝（图13）；散开后再翻炒几下，见鸡肉变色、已有8分熟（图14）即可盛出，沥干油渍。

　　初学者如果不会掌握油的温度，可以在下锅前加入1～2大匙冷油和鸡肉拌匀，以免鸡丝或鸡粒粘在一起。

图13

图14

过水汆烫：
　　为减少用油量，也可以用多量的热水汆烫鸡肉之后再炒，鸡肉下锅之后立刻改用中小火浸泡至9分熟，捞出、备炒。
　　要烫的鸡肉在腌的时候可以多加一点点淀粉，以保持鸡肉的滑嫩，因为水温不如油温高，会使淀粉流失，但是淀粉也不能多用，以免在鸡肉的表面凝结成小块，炒的时候反而会脱落、糊化，影响菜肴的口感与美观。

Cooking tips:
Pre-stir-fry with oil:

In Chinese cooking, chicken meat used in stir-fry dishes is traditionally pre-cooked in excess oil until it is almost done. This additional step helps to keep the chicken meat juicy and tender, and it also improves the flavor. Other ingredients, spices, and seasonings are then added after this pre-stir-fry procedure.

Pre-stir-frying requires more oil than the meat. Heat the oil to medium high (120~160℃)——120~140℃ for chicken strings, and 140~160℃ for chicken slices and cubes. Add in the meat, separate chicken strings with chopsticks quickly (pic13) and stir-fry until it is almost cooked (pic14). Remove the chicken strings and drain the oil. If you are not familiar with the temperature of oil, you may mix in 1 to 2 Tbsp of cold oil with the meat prior to adding it into the hot oil. In this way you can make sure that the chicken meat won't stick in hot oil.

Blanch:

You may boil the chicken meat instead of pre-stir-frying to reduce the fat. If boiling, the amount of cornstarch used in marinating should be increased slightly because some may be lost during boiling. However, too much cornstarch will damage the texture and appearance of your dish.

香根银芽炒鸡丝
Stir-fried Chicken Shreds with Bean Sprouts

材料：
鸡胸肉1片（约150克）、绿豆芽200克、葱丝1大匙、香菜3根
调味料：
（1）盐 1/4 茶匙、水1大匙、淀粉1茶匙
（2）酒1茶匙、盐 1/4 茶匙、水2～3大匙、淀粉水少许、麻油少许
做法：
1. 鸡胸肉修整好，顺纹切成约5厘米长的丝。用调味料（1）轻轻抓拌均匀，腌20～30分钟。
2. 绿豆芽摘去头尾成银芽，洗净、沥干；香菜切成短段。
3. 依照第60页方法将鸡丝过油或汆烫，待鸡肉至8～9分熟时，捞出。
4. 锅中热1大匙油，放下葱丝和银芽，炒到银芽快熟时，淋下酒增香、加盐调味炒匀。
5. 把鸡丝放入锅中，再由锅边淋下水，大火炒匀。
6. 以淀粉水勾上薄芡，加入香菜段，关火，再略加搅拌即可装盘。

Ingredients:
150g. chicken breast, 200g. mung bean sprouts, 1 tbsp shredded green onion, 3 stalks cilantro

Seasonings:
(1) 1/4 tsp salt, 1 tbsp water, 1 tsp cornstarch
(2) 1 tsp wine, 1/4 tsp salt, 2～3 tbsp water, a little of cornstarch paste, sesame oil

Procedures:
1. Trim chicken breast. Shred it finely into 5cm long. Gently mix with seasonings (1), marinate for 20~30 minutes.
2. Trim bean sprouts, rinse and drain dry; cut cilantro to sections.
3. According to P.62 to stir-fry or blanch the chicken shreds to almost cooked. Drain.
4. Heat 1 tablespoon of oil to stir-fry green onion and bean sprouts. Splash wine when bean sprouts are just cooked. Season with salt.
5. Return chicken to wok, sprinkle water in, stir fry over high heat until evenly mixed.
6. Thicken with a little of cornstarch paste, add cilantro at last. Turn off the heat, mix and transfer to a serving plate.

细说鸡胸

炒鸡丝拉皮
Stir-fried Chicken Shreds Salad

材料：
鸡胸肉约120克、新鲜粉皮1包、小黄瓜1条、葱1根

调味料：
（1）盐少许、水1～2大匙、淀粉1茶匙
（2）芝麻酱2大匙、温水3大匙、芥末酱或芥末粉1大匙、醋1茶匙、淡色酱油2大匙、麻油1大匙、盐少许、糖 1/4 茶匙

做法：
1. 鸡胸肉切成细丝，用调味料（1）拌均匀，腌20～30分钟。
2. 小黄瓜切成细丝；粉皮切成宽条，冲洗一下，沥干水分，和黄瓜分别排在盘中；葱斜切成细丝。
3. 芝麻酱先加温水调软，再陆续加入其余的调味料（2）调匀备用（如用芥末粉则须先用温水调稀再加入）。
4. 将1杯油加热至130℃左右，放下鸡丝过油，待鸡丝熟了即用漏勺捞出。
5. 趁热把葱丝放在鸡丝上拌一下，再放在粉皮上，淋下调味汁，上桌后拌匀即可。

※若用干粉皮则需要先泡水，再放入滚水中把它烫软，冲凉后再来拌。

Ingredients:
120g. chicken breast, 1 piece mung bean sheet,
1 cucumber, 1 stalk green onion

Seasonings:
(1) $1/6$ tsp salt, 1 ~ 2 tbsp water, 1 tsp cornstarch
(2) 2 tbsp sesame paste, 3 tbsp warm water,
   1 tbsp mustard powder or wasabi, 2 tbsp soy sauce,
   1 tsp vinegar, 1 tbsp sesame oil, salt to taste, $1/4$ tsp sugar

Procedures:
1. Trim chicken breast, then shred it. Marinate with seasonings (1) for 20 ~ 30 minutes.
2. Shred cucumber; cut mung bean sheet into strings, rinse and drain, arrange on a plate with cucumber. Shred green onion.
3. Dissolve sesame paste with water, then add other seasonings (2), mix evenly.
4. Heat 1 cup of oil to 130℃, run chicken through oil, as soon as the chicken is cooked, drain it with a sifter.
5. Add green onion to chicken while chicken is hot, mix them to enhance the fragrant. Remove and place on top of bean sheets. Pour sauce over chicken. Mix evenly before eat.

※You may use dried mung bean sheet to make this dish. Soak it until soft first, then boil for a while and rinse to cool.

豌豆炒鸡丝
Chicken Shreds with Baby Snow Peas

材料：
鸡胸肉150克、
新鲜豌豆150克或冷冻青豆 1 1/2 杯、
葱末 1/2 大匙、熟火腿丝1大匙

调味料：
（1）盐 1/4 茶匙、水1大匙、淀粉 1/2 茶匙、蛋白 1/2 大匙
（2）酒1茶匙、盐 1/3 茶匙、水 1/2 杯、淀粉水1茶匙、麻油数滴

做法：
1. 鸡胸肉修整好后切成细丝，用调味料（1）仔细拌匀，腌约1小时。
2. 新鲜豌豆用滚水汆烫一下，即刻捞出后用冷水冲凉；如用冷冻青豆，烫一下就可以用了。
3. 锅中把1杯油加热至140℃，放下鸡丝过油炒至8～9分熟，捞出后把油倒出，仅留1大匙左右。
4. 放下葱末爆香后，将豌豆粒及鸡丝一起下锅，淋下酒、盐和水。煮滚后用淀粉水勾上薄芡，再淋入数滴麻油，炒匀。装盘后撒下火腿丝即可。

Ingredients:
150g. chicken breast,
150g. fresh baby snow peas or
1 1/2 cups frozen snow peas,
1/2 tbsp chopped green onion,
1 tbsp cooked Chinese ham shreds

Seasonings:
(1) 1/4 tsp salt, 1 tbsp water, 1/2 tsp cornstarch,
  1/2 tbsp egg white
(2) 1 tsp wine, 1/3 tsp salt, 1/2 cup water,
  1 tsp cornstarch paste, few drops of sesame oil

Procedures:
1. Trim chicken, cut into fine shreds, marinate with seasonings (1) for 1 hour.
2. Blanch baby snow peas, drain and rinse to cool. If you use the frozen snow peas, just blanch and drain it.
3. Heat 1 cup of oil to 140℃, run chicken shreds through oil quickly, separate them from stick together with chopsticks, drain them when the color turn white. Pour oil away.
4. Keep only 1 tablespoon of oil to saute green onion. Add chicken and snow peas, sprinkle wine, salt and water, stir evenly and bring to a boil. Thicken with cornstarch paste, drizzle sesame oil and then transfer to a serving plate, add ham shreds on top, serve hot.

鸡丝炒牛蒡
Stir-fried Chicken with Bunduck

材料：
牛蒡 1/2 枝、鸡胸肉150克、
葱1根（切段）、芝麻1大匙、麻油1茶匙

调味料：
香菇酱油1 1/2 大匙、酒 1/2 大匙、
味霖（日式调味米酒）1大匙、糖 1/2 茶匙、水3大匙

做法：
1. 牛蒡用刀背刮去外皮后切成细丝，浸在冷水中以防变色（水中可加醋1茶匙），取出后沥干水分。
2. 鸡肉切丝，锅中烧热油3大匙，放入葱段和鸡丝，炒熟盛出。
3. 放下牛蒡丝再炒，炒至颜色变深时，加入调味料，改成小火，再炒2～3分钟，炒至汤汁几乎完全收干。
4. 加入鸡丝炒匀，再滴下麻油，撒下炒香的芝麻即可。

※牛蒡是属于硬性食材，因此鸡丝不用腌。

Ingredients:
1/2 bunduck, 150g. chicken breast,
1 stalk green onion (sectioned),
1 tbsp sesame seeds, 1 tsp sesame oil

Seasonings:
1 1/2 tbsp soy sauce, 1/2 tbsp wine,
1 tbsp mirin, 1/2 tsp sugar, 3 tbsp water

Procedures:
1. Peel off skin from bunduck and then shred it. Soak in water (add 1 teaspoon of vinegar in water) to keep away from getting dark. Drain before using.
2. Shred chicken. Sauté chicken and green onion sections with 2 ~ 3 tablespoons of oil. Remove when cooked.
3. Add bunduck in, continue to stir-fry, when the color of bunduck turns deeper, add seasonings and then turn to low heat, continue to stir-fry for 2 ~ 3 minutes until water is almost absorbed.
4. Add chicken in, stir-fry evenly. Add sesame seeds and drizzle sesame oil at last.

※Since bunduck is hard in texture, so we don't have to marinate chicken.

Chicken Shreds with Pickled Mustard Green

雪菜百叶烩鸡丝
Chicken Shreds with Pickled Mustard Green

材料：
鸡胸肉120克、雪里红150克、百叶 1/2 叠
（5张）或新鲜豆皮1片、葱1根（切段）、清汤 2/3 杯、小苏打 1/2 茶匙

调味料：
（1）盐 1/6 茶匙、水1大匙、淀粉 1/2 茶匙、蛋白 1/2 大匙
（2）酱油1茶匙、盐 1/3 茶匙、淀粉水1茶匙、麻油数滴

做法：
1. 鸡胸肉切丝，用调味料（1）拌匀，腌1小时。
2. 雪里红冲洗干净后切小丁，挤干水分（尾端老叶不要）。
3. 煮滚3杯水，加入小苏打，关火。百叶切成3厘米的宽条，放入苏打水中浸泡。见百叶的颜色变白且变软时，便可将百叶取出，放在清水中多漂洗几次，沥干水分。如果用新鲜豆皮，切成宽条、撕散开来即可使用。
4. 把1杯油烧至7分热，过油炒熟鸡丝或用滚水烫熟鸡丝，捞出。
5. 另用1大匙油炒香葱段，加入雪里红炒一下，再加入酱油烹香略炒一下。加入百叶、清汤和盐炒匀，一起煮滚。
6. 当一煮滚时立即放下鸡丝拌炒，再煮滚即可用淀粉水勾芡，滴下麻油便可盛出。

Ingredients:
120g. chicken breast, 150 g. pickled mustard green,
1/2 pile dried bean curd sheet (5 pieces) or 1 fresh tofu pack,
1 stalk green onion (sections), 2/3 cup soup stock, 1/2 tsp baking soda

Seasonings:
(1) 1/6 tsp salt, 1 tbsp water, 1/2 tsp cornstarch, 1/2 tbsp egg white
(2) 1 tsp soy sauce, 1/3 tsp salt, 1 tsp cornstarch paste, few drops of sesame oil

Procedures:
1. Shred chicken breast and marinate it with seasonings (1) for 1 hour.
2. Rinse mustard green thoroughly, chop and squeeze out the liquid (discard rough leaves).
3. Bring 3 cups of water to a boil, add baking soda in, turn off the heat. Cut bean curd sheet to 3cm wide, soak into soda water. Remove when bean curd sheet turn soft and the color turn lighter. Rinse with clear water gently, remove and drain. For the fresh tofu pack, just cut it to wide strips.
4. Heat 1 cup of oil to stir-fry chicken to almost done, or you may blanch it, remove and drain.
5. Heat 1 tablespoon of oil to sauté green onion sections, add pickled mustard green, stir-fry briefly. Add soy sauce to enhance the fragrant. Add bean curd sheet, soup stock and salt, bring to a boil.
6. Add chicken when boiling. Thicken with cornstarch paste when boils again. Drizzle sesame oil at last. Remove.

Chicken and Dried Scallop Potage

瑶柱鸡丝羹
Chicken and Dried Scallop Potage

材料：
鸡胸肉120克、干贝3粒、笋子1棵、
葱1根、姜2片、鸡清汤6杯、香菜适量
调味料：
（1）盐 $1/3$ 茶匙、水1大匙、淀粉 $1/2$ 大匙
（2）酒1大匙、盐1茶匙、淀粉水2大匙、白胡椒粉适量、麻油数滴
做法：
1. 鸡胸肉切成细丝，用调味料（1）拌匀，腌30分钟以上。
2. 干贝加水（水量要超过干贝1~2厘米），放入电锅中，蒸30分钟至软，放凉后撕散成细丝。
3. 笋子削好，切成丝后放入碗中，加水盖过笋丝，也蒸20分钟至熟。
4. 起油锅用1大匙油爆香葱段和姜片，待葱姜略焦黄时，淋下酒和清汤，再放入干贝（连汁）和笋丝（连汁）一起煮滚，夹除葱姜。
5. 加入鸡丝，一边放一边轻轻地搅动使鸡丝散开，加盐调味，再以淀粉水勾芡。
6. 最后加入胡椒粉和麻油增香，装碗后加入香菜段。

※鸡丝放入汤中之前，可以先用筷子挑散，或加少许的水调散开。

Ingredients:
120g. chicken breast, 3 dried scallops, 1 bamboo shoot, 1 stalk green onion (sectioned),
2 slices ginger, 6 cups chicken soup stock, 1 stalk cilantro (sectioned)
Seasonings:
(1) $1/3$ tsp salt, 1 tbsp water, $1/2$ tbsp cornstarch
(2) 1 tbsp wine, 1 tsp salt, 2 tbsp cornstarch paste, a little of white pepper and sesame oil
Procedures:
1. Shred chicken and then marinate with seasonings (1) for 30 minutes.
2. Steam scallops with water (the water should cover scallops at least for 1 ~ 2cm deep) for 30 minutes until soft. Remove and let cool. Tear it apart.
3. Trim bamboo shoot, shred it and then steam with water for 20 minutes.
4. Heat 1 tablespoon of oil to fry green onion sections and ginger, pour wine and soup stock in, also add scallop and bamboo shoot (together with both steamed liquid), bring to boil, discard green onion and ginger.
5. Add chicken shreds in, stir gently while adding to prevent sticking together. Season with salt and then thicken with cornstarch paste.
6. Enhance the fragrant with pepper and sesame oil at last. Add cilantro sections on top after remove to a bowl.

※Separate chicken shreds with chopsticks before adding to the soup or add some water to chicken to avoid from sticking together.

Fried Chicken with Lemon Sauce

柠檬鸡片
Fried Chicken with Lemon Sauce

材料：
鸡胸肉2片、洋葱丝 1/2 杯、蒜末1茶匙、柠檬2个、鸡蛋1个（打散）、玉米粉4～5大匙

调味料：
（1）盐1茶匙、酒2茶匙、白胡椒粉少许、水 1/2 杯
（2）糖4大匙、水 2/3 杯、盐 1/3 茶匙、淀粉1茶匙、淡色酱油1茶匙

做法：
1. 把鸡胸肉内侧的小里脊肉拉掉，修整掉软骨，在两面轻轻地剁上一些刀口。加入调味料（1）搅匀，腌15分钟。
2. 把1个半的柠檬挤汁，有3～4大匙的量，磨下一点柠檬皮末，加入调味料（2）中一起调匀；另外半个切片做盘饰。
3. 鸡片先蘸蛋汁，再蘸上玉米粉，轻轻压一下，放1～2分钟。
4. 炸油烧至8分热，加入鸡肉，用中小火炸约1.5分钟，捞出鸡肉。
5. 油再烧热，放入鸡肉用大火再炸一次，炸至外皮金黄，捞出，沥干油，切成长块装盘。
6. 另用1大匙油炒香洋葱丝和蒜末，倒下柠檬调味料煮滚成芡汁，调整一下味道后，淋到鸡片上。

Ingredients:
2 pieces chicken breast, 1/2 cup onion shreds,
1 tsp chopped garlic, 2 lemons, 1 egg (beaten), 4～5 tbsp cornstarch

Seasonings:
(1) 1 tsp salt, 2 tsp wine, pinch of white pepper, 1/2 cup water
(2) 4 tbsp sugar, 2/3 cup water, 1/3 tsp salt, 1 tsp cornstarch, 1 tsp light colored soy sauce

Procedures:
1. Trim off the chicken tender from chicken breast. Cut some lines on both sides. Mix with seasonings (1), marinate for 15 minutes.
2. Squeeze juice from 1 1/2 lemons to get about 3～4 tablespoons of lemon juice. Grate some lemon zest, mix together with seasonings (2). Slice the remaining half lemon for decoration.
3. Dip chicken breast in beaten egg mixture, then coat with cornstarch, press the chicken gently and set aside for 1～2 minutes.
4. Heat the deep-frying oil to 160℃, deep-fry chicken for 1.5 minutes over medium-low heat. Remove chicken.
5. Reheat oil, deep-fry chicken over high heat until the surface becomes golden brown. Remove and drain off the oil. Cut into pieces, place on a plate.
6. Sauté onion and garlic with 1 tablespoon of oil, pour seasonings (2) in, bring to a boil. Taste the sauce and season again if needed, pour sauce over chicken. Serve hot.

芥汁鸡片
Pouched Chicken with Mustard Sauce

材料：
鸡胸肉250克、小油菜6棵、大蒜泥1茶匙
调味料：
（1）盐 1/3 茶匙、水1大匙、淀粉2茶匙、蛋白1大匙
（2）美奶滋2大匙、法式黄色芥末酱1茶匙、淡色酱油2茶匙、糖 1/2 茶匙、柠檬汁或白醋 1/2 茶匙、橄榄油 1/2 大匙

做法：
1. 鸡胸肉处理好，冷冻较硬后切成薄片，用调味料（1）拌匀，冷藏1小时。
2. 小油菜摘好；将调味料（2）和蒜泥在碗中调匀。
3. 烧开4～5杯水，加1茶匙盐和1大匙油在水中，先烫小油菜，烫熟后捞出，排入盘中。
4. 将水再烧开，放入鸡片，改成小火，轻轻推散鸡片，泡煮至鸡片已熟，捞出，沥干水分。
5. 把鸡片放在盘中，淋下芥末调味汁即可。

Ingredients:
250g. chicken breast,
6 stalks green cabbage,
1 tsp mashed garlic

Seasonings:
(1) 1/3 tsp salt, 1 tbsp water,
   2 tsp cornstarch, 1 tbsp egg white
(2) 2 tbsp mayonnaise, 1 tsp yellow French mustard,
   2 tsp light colored soy sauce, 1/2 tsp sugar,
   1/2 tsp lemon juice or vinegar, 1/2 tbsp olive oil

Procedures:
1. Trim chicken breast, frozen until hard enough to slice it. Mix with seasonings (1) for 1 hour.
2. Trim green cabbage; combine all seasonings (2) with mashed garlic.
3. Bring 4～5 cups of water to a boil, add 1 teaspoon of salt and 1 tablespoon of oil in water. Blanch green cabbage until cooked. Drain and then arrange on a plate.
4. Bring the water to a boil again, add chicken in. Turn to low heat, stir gently to separate chicken slices. Cook until done. Drain and place on the plate. Pour the sauce over chicken, serve hot.

韩风炒鸡肉
Stir-fried Chicken, Korean Style

材料：
鸡胸肉250克、洋葱 1/2 个、干香菇3～4朵、葱3～4根

调味料：
酱油3大匙、蒜末1大匙、姜汁1茶匙、胡椒粉 1/2 茶匙、
糖1大匙、麻油1大匙、白芝麻1大匙、水3大匙

做法：
1. 鸡肉放冷冻室中冻硬一些，取出、切成薄片。
2. 香菇泡软后剪掉菇蒂，切成宽条；洋葱也切宽条；葱切斜片。
3. 调味料在碗中先调匀，再将鸡肉、香菇、洋葱和葱都放入，腌10分钟使材料入味。
4. 用大一点的平底厚锅，加热后把一半量调好味道的鸡肉和蔬菜放入，大火炒熟，盛出，再炒另一半，装入盘中。

※一次炒的量少一点，才能把肉炒香且炒的嫩。

Ingredients:
250g. chicken breast, 1/2 onion,
3～4 dried shitake mushrooms,
3～4 stalks green onion

Seasonings:
3 tbsp soy sauce, 1 tbsp chopped garlic,
1 tsp ginger juice, 1/2 tsp pepper, 1 tbsp sugar,
1 tbsp sesame oil, 1 tbsp sesame seeds, 3 tbsp water

Procedures:
1. Slice chicken breast, it will be easier to slice the meat if you frozen it for 2 hours.
2. Soak dried shitake mushrooms to soft, cut to stripes; cut onion to stripes; section green onions.
3. Mix seasonings in a large bowl, marinate chicken, shitake mushrooms, onion and green onion for 10 minutes.
4. Heat a large heavy pan to very hot, fry half amount of chicken and vegetables over high heat until done. Remove and fry the remaining half again.

※Don't stir fry too much chicken at one time to get better texture and keep the tenderness of meat.

Chicken Salad with Miso Sauce

味噌酱拌鸡柳
Chicken Salad with Miso Sauce

材料：
鸡胸肉200克、金针菇1包（约120克）、
洋葱 1/2 个、红辣椒1个、炒过的白芝麻1大匙

调味料：
（1）盐 1/3 茶匙、水2大匙、淀粉 1/2 大匙
（2）味噌酱1 1/2 大匙、味霖1大匙、糖1/4茶匙、麻油1茶匙、橄榄油 1/2 大匙、水3大匙

做法：
1. 鸡胸肉修整好后切成粗条，用调味料（1）拌匀，腌20～30分钟以上。
2. 洋葱切丝后泡入冰水中，约10分钟后沥干水分，放入盘中。
3. 金针菇切除根部、冲洗一下；红辣椒去籽、切丝。
4. 调味料（2）先调好，在小锅中煮滚后立刻关火。
5. 烧开4杯水，放下金针菇烫煮一滚，捞出，沥干水分，放在碗中。
6. 水再烧开，放入鸡丝，用筷子搅散鸡肉，待鸡肉煮熟后捞出，也放入碗中，加上红辣椒丝和味噌酱拌匀，盛放在洋葱丝上，撒上芝麻。

※不同品牌的味噌的咸度会不相同，炒好后要先尝一下味道，调整咸度。

Ingredients:
200g. chicken breast, 1 pack needle mushroom (about 120g.),
1/2 onion, 1 red chili, 1 tbsp white sesame seeds

Seasonings:
(1) 1/3 tsp salt, 2 tbsp water, 1/2 tbsp cornstarch
(2) 1 1/2 tbsp miso, 1 tbsp mirin, 1/4 tsp sugar,
   1 tsp sesame oil, 1/2 tbsp olive oil, 3 tbsp water

Procedures:
1. Trim chicken breast, cut into strips, marinate with seasonings (1) for 20 ~ 30 minutes.
2. Cut onion to thin shreds, soak in ice water for 10 minutes, drain and arrange on a plate.
3. Trim needle mushroom, rinse and drain dry; remove seeds from red chili, then shred it.
4. Mix seasonings (2) in a bowl, bring to a boil in a sauce pan.
5. Boil 4 cups of water, blanch needle mushroom first, remove when cooked. Drain dry and place in a large bowl.
6. Boil water again, put chicken in. Stir chicken gently, remove to bowl when cooked. Mix with red chili and miso sauce, transfer to a plate. Sprinkle white sesame seeds over chicken at last.

※Different brand of miso taste quite differently, try the taste then adjust it.

Stir-fried Chicken with Fermented Tofu Sauce

鸿喜腐乳鸡
Stir-fried Chicken with Fermented Tofu Sauce

材料：
鸡胸肉200克、鸿喜菇1包、
葱1根、姜6～7小片、大蒜5～6片

调味料：
（1）盐 $1/4$ 茶匙、水1大匙、淀粉 $1/2$ 茶匙
（2）豆腐乳1大匙、腐乳汁1大匙、糖1茶匙、麻油 $1/3$ 茶匙

做法：
1. 鸡胸肉切成较粗的丝条，用调味料（1）拌匀，腌20分钟。
2. 豆腐乳压碎成泥状，加入腐乳汁、糖和2大匙水调匀。
3. 鸿喜菇（又叫灵芝菇）切去根部，大略的分散开一些。
4. 锅烧热，放下1大匙的油，先把鸿喜菇以中火加以煸炒至有焦痕，且有香气时盛出。再加入2大匙油把鸡丝炒熟，也盛出。
5. 放下姜和蒜片继续炒香，再加入腐乳汁，倒回鸡丝、鸿喜菇和葱段，以中大火快速炒匀，滴下麻油便可起锅。

※喜欢吃辣的话可以选用麻辣豆腐乳，或加些辣油同炒。

Ingredients:
200g. chicken breast, 1 pack Hong-shi mushrooms,
1 stalk green onion, 6 ~ 7 slices ginger, 5 ~ 6 slices garlic

Seasonings:
(1) $1/4$ tsp salt, 1 tbsp water, $1/2$ tsp cornstarch
(2) 1 tbsp fermented bean curd (tofu), 1 tbsp sauce from fermented tofu,
    1 tsp sugar, 1/3 tsp sesame oil

Procedures:
1. Shred chicken breast, marinate with seasonings (1) for 20 minutes.
2. Smash fermented bean curd, mix evenly with the sauce, sugar and 2 tablepoons of water.
3. Trim Hong-shi mushroom, separate them a little.
4. Stir-fry mushroom with 1 tablespoon of oil until brown. Remove. Add 2 tablespoons of oil to stir-fry chicken shreds until done, remove.
5. Add ginger and garlic, fry until fragrant. Add mixed seasonings in, return chicken, mushroom and green onion, stir-fry over medium-high heat until sauce becomes thicker. Drizzle sesame oil at last.

※You may choose the spicy fermented tofu to make this dish or add some red chili oil if you prefer spicy dish.

Diced Chicken with Peppers

辣子鸡丁
Diced Chicken with Peppers

材料：
鸡胸肉250克、青椒1个、茭白笋2棵、红辣椒2个、葱1根、姜8～10小片

调味料：
（1）酱油1大匙、淀粉2茶匙、水1大匙
（2）辣椒酱 $1/2$ 大匙、酱油$1\, 1/2$ 大匙、醋1大匙、糖 $1/2$ 茶匙、盐 $1/4$ 茶匙、水4大匙、淀粉1茶匙、麻油 $1/4$ 茶匙、胡椒粉少许

做法：
1. 鸡肉用刀先轻轻拍松，再切成1.5厘米大小的丁，用调味料（1）拌匀，腌至少半小时。
2. 青、红辣椒洗净、去籽，和茭白笋一起切成与鸡肉同样大小的方丁；葱切小段。
3. 碗中先将调味料（2）调匀备用。
4. 将2杯油烧到8分热后，倒下腌过的鸡丁，泡炸15～20秒钟至鸡丁转白即可捞出，余油倒出。
5. 另在锅内烧热2大匙油，爆炒葱段、姜片和茭白笋，片刻之后，再放入青、红辣椒丁，翻炒数下至产生香气。
6. 倒下鸡丁，同时淋下调匀的调味料（2），用大火迅速拌炒均匀即可盛出。

Ingredients:
250g. chicken breast, 1 green pepper, 2 jiao-bai bamboo shoots, 2 red chilies, 1 stalk green onion, 8 ~ 10 slices ginger

Seasonings:
(1) 1 tbsp soy sauce, 2 tsp cornstarch, 1tbsp water
(2) $1/2$ tbsp hot chili paste, $1\, 1/2$ tbsp soy sauce, 1 tbsp brown vinegar, $1/2$ tsp sugar, $1/4$ tsp salt, 4 tbsp water, 1 tsp cornstarch, 1/4 tsp sesame oil, a pinch of pepper

Procedures:
1. Cut the chicken into 1.5cm cubes. Marinate with seasonings (1) for at least 30 minutes.
2. Dice green pepper, red peppers, jiao-bai bamboo shoot, and green onion to the similar size as chicken.
3. Mix the seasonings (2) in a bowl.
4. Heat 2 cups of oil to 160℃. Fry the diced chicken for about 15 ~ 20 seconds. Remove and drain off the oil.
5. Heat another 2 tablespoons of oil to stir fry the green onion, ginger, and bamboo shoot first, then add green pepper and red pepper, stir-fry for about 10 seconds until fragrant.
6. Add chicken and the seasoning sauce. Sauté over high heat until sauce is thickened and heated thoroughly. Serve hot.

Stir-fried Chicken with Bean Paste

川香鸡片
Stir-fried Chicken with Bean Paste

材料：
鸡胸肉300克、圆白菜300克、
豆腐干5片、红辣椒2个、青蒜 1/2 根

调味料：
（1）酱油1大匙、淀粉 1/2 大匙、水1大匙、蛋白1大匙
（2）甜面酱1 1/2 大匙、辣豆瓣酱1大匙、酱油 1/2 大匙、水1大匙、糖2茶匙

做法：
1. 鸡肉打斜切片，用调味料（1）拌匀，最好腌1小时以上。
2. 圆白菜切成大片；豆腐干斜刀片成薄片；红辣椒去籽、切片；青蒜切丝备用。
3. 甜面酱等调味料（2）在小碗内先调匀备用。
4. 锅内烧热1杯油，将鸡肉放入快炒，约8分熟时盛出，油倒出。
5. 用1大匙油来炒圆白菜及豆腐干，加少许清水将圆白菜炒软，盛出。
6. 另用2大匙油炒香甜面酱料，10～15秒钟至有香气透出。
7. 再将鸡肉、红辣椒和圆白菜等倒回锅内，以大火拌炒均匀，放下青蒜再炒一下即可。

※炒甜面酱料时要用小火来炒才会产生香气，而且避免有焦苦味。

Ingredients:
300g. chicken breast, 300g. cabbage, 5 pieces dried tofu, 2 red chilies, 1/2 green garlic

Seasonings:
(1) 1 tbsp soy sauce, 1/2 tbsp cornstarch, 1 tbsp water, 1 tbsp egg white
(2) 1 1/2 tbsp sweet soybean paste, 1 tbsp hot bean paste,
    1/2 tbsp soy sauce, 1 tbsp water, 2 tsp sugar

Procedures:
1. Slice chicken meat, marinate with seasonings (1) for at least 1 hour.
2. Cut the cabbage into large pieces; slice dried tofu; deseed and then slice red chilies; shred green garlic.
3. Mix the seasonings (2) in a bowl.
4. Heat 1 cup of oil to stir-fry chicken until almost done. Remove chicken and pour away the oil.
5. Stir-fry cabbage and dried tofu with 1 tablespoon of oil, add a little of water, cook until the cabbage softens. Remove.
6. Heat 2 tablespoons of oil to stir-fry the sauce for about 10～15 seconds until fragrant.
7. Return chicken, red chili and cabbage to sauce, stir-fry thoroughly. Add green garlic at last, mix evenly, serve.

※You must stir-fry the sauce over low heat to make it fragrant and avoid from burning.

墨西哥鸡肉法西达斯
Chicken Fajitas

材料：
鸡胸肉250克、洋葱1/2个、大蒜片3~4片、青椒1/2个、绿辣椒2个、墨西哥饼或口袋饼4张、巧达起司（乳酪）4大匙或巧达起司片3片、酸奶油和酪梨酱各1杯

调味料：
盐、胡椒适量调味，酒 1/2 大匙

做法：
1. 鸡胸肉切条片状，撒少许盐和胡椒粉调味；洋葱切丝；青椒和绿辣椒也切丝备用。
2. 锅中用2大匙油炒香洋葱和大蒜片，再放下鸡肉炒散，炒到鸡肉略有焦黄色且有香气，加入青椒和绿辣椒，大火炒香。再淋下酒和一点水炒匀，盛出。
3. 墨西哥饼烘烤热之后，放上炒鸡肉，再撒上巧达起司（可以把起司片切成丝再用），随个人喜爱淋上酸奶油和酪梨酱包食。

酪梨酱：
1. 熟酪梨（又叫牛油果）2个，切半、去籽，挖出酪梨肉放入碗中，先和柠檬汁（约2大匙）拌一下，沥出柠檬汁。加入盐和少许小茴香调味、压碎、拌匀。
2. 另外准备2大匙洋葱细末、1/2茶匙蒜泥、香菜剁碎（约1茶匙）、2大匙剁碎的番茄和酪梨肉一起混合均匀。再淋下沥出的柠檬汁（约1大匙）。放在室温1小时后食用。

Ingredients:
250g. chicken breast, 1/2 onion, 3 ~ 4 slices garlic, 1/2 green pepper, 2 green chilies
4 pieces tortilla, 4 tbsp cheddar cheese or 3 pieces cheddar cheese,
1 cup sour cream, 1 cup guacamole

Seasonings:
salt and pepper to taste, 1/2 tbsp wine

Procedures:
1. Slice chicken to strips, sprinkle salt and pepper to taste. Shred onion, green pepper and green chilies.
2. Heat 2 tablespoons of oil to sauté onion and garlic, add chicken, stir-fry until the chicken is light browned and cooked. Add green pepper and chilies in, stir-fry over high heat. Splash wine and a little of water in, stir-fry evenly.
3. Reheat tortilla, put chicken on top, sprinkle cheddar cheese on (shred cheddar cheese first if you use the whole pieces), then add sour cream and guacamole as your choice.

Guacamole:
1. 2 ripe avocados, halved, seeded and scoop the pulp to a large bowl. Add 2 tablespoons of lemon juice, toss to coat. Drain, reserve the juice. Season with salt and cumin, then mash it.
2. Fold in 2 tablespoons of chopped onion, 2 tablespoons of chopped tomato, 1 teaspoon of cilantro and 1/2 teaspoon of mashed garlic. Add 1 tablespoon of reserved lemon juice. Let it sit at room temperature for 1 hour before serving.

Fried Chicken with Curry Sauce

咖喱鸡排
Fried Chicken with Curry Sauce

材料：
鸡胸肉2片、圆白菜丝2杯、
面粉 1/2 杯、鸡蛋1个（打散）、面包粉1杯、
洋葱丁 1/2 杯、大蒜末1茶匙、冷冻三色蔬菜1杯

调味料：
（1）盐 1/2 茶匙、胡椒粉 1/4 茶匙、酒1大匙、水 1/2 杯
（2）清汤或水1杯、咖喱块2小块、盐适量

做法：
1. 鸡胸剔去软骨和筋之后，在鸡胸肉上剁一些刀口，加入调味料（1），腌20~30分钟。
2. 圆白菜丝泡入冰水中5~10分钟，沥干，并以纸巾吸干水分。
3. 将鸡肉先蘸上面粉，再在蛋汁中蘸一下，最后蘸满面包粉。
4. 把4杯炸油烧至7分热，放入鸡排，以小火炸约2.5分钟，捞出。
5. 把油再烧至8分热，放下鸡排，以大火再炸20秒钟，至鸡排成为金黄色且酥脆。捞出鸡排、沥干油渍，切成片，排放入盘中。
6. 在炸鸡排之时，另用1大匙油炒香洋葱丁和大蒜末，再加入三色蔬菜同炒，倒下清汤并加入咖喱块，一起煮至咖喱块溶化。再适量调味后盛出，和鸡排一起上桌。

Ingredients:
2 pieces chicken breast, 2 cups cabbage shreds, 1/2 cup flour, 1 egg (beaten), 1 cup bread crumbs, 1/2 cup diced onion, 1 tsp chopped garlic, 1 cup frozen mixed vegetables

Seasonings:
(1) 1/2 tsp salt, 1/4 tsp pepper, 1 tbsp wine, 1/2 cup water
(2) 1 cup soup stock or water, 2 small curry cubes, salt to taste

Procedures:
1. Trim chicken breast, chop some cuts on chicken, brine with seasonings (1) for 20 ~ 30 minutes.
2. Soak cabbage shreds in ice water for 5 ~ 10 minutes, drain and then dry it with paper towel.
3. Cover chicken with flour, then dip in egg mixture, and finally coat with bread crumbs.
4. Heat 4 cups of oil to 140℃, deep fry chicken over low heat for about 2.5 minutes. Remove chicken.
5. Reheat oil to 160℃, deep fry chicken over high heat until golden brown, about 20 seconds. Drain and cut to pieces, arrange on a plate.
6. Sauté onion and garlic with 1 tablespoon of oil, add mixed vegetables, stir fry for a while. Add stock and curry cubes in, cook until curry melt. Season again if needed. Serve with fried chicken.

红莓鸡片
Fried Chicken with Cranberry Sauce

材料：
鸡胸肉300克、蔓越莓3大匙、番薯粉 2/3 杯

调味料：
（1）盐 1/3 茶匙、酒2茶匙、白胡椒粉1/4茶匙、鸡蛋1个
（2）水1杯、柠檬汁1大匙、盐 1/4 茶匙、淀粉1 1/2 茶匙

做法：
1. 将鸡胸肉打斜切成片，用调味料（1）拌匀，腌15～20分钟。再蘸上番薯粉，放置2～3分钟。
2. 蔓越莓（一种生长在北美的植物）加1杯水先泡10分钟，连水用果汁机打一下，不要打得太细，仍保有些果粒，再和其他的调味料（2）调匀。
3. 将3杯炸油烧至8分热（160℃），放入鸡片，用中火炸40～50秒钟，捞出。
4. 油再烧热，放入鸡片，以大火再炸10秒钟，捞出，沥干油后盛入盘中。
5. 炸油倒出，另用1大匙油炒煮蔓越莓调味汁，煮滚立即关火，将汁淋到鸡片上，尽快上桌，趁热食用。

Ingredients:
300g. chicken breast,
3 tbsp dried cranberries,
2/3 cup sweet potato powder

Seasonings:
(1) 1/3 tsp salt, 2 tsp wine,
   1/4 tsp pepper, 1 egg
(2) 1 cup water, 1 tbsp lemon juice,
   1/4 tsp salt, 1 1/2 tsp cornstarch

Procedures:
1. Slice chicken, mix evenly with seasonings (1), marinate for 15 ~ 20 minutes. Coat with sweet potato powder, leave for 2 ~ 3 minutes.
2. Soak cranberries with 1 cup of water for 10 minutes. Blend with water, do not blend it too fine. Mix with other seasonings (2).
3. Heat 3 cups of oil to 160℃, deep fry chicken for 40 ~ 50 seconds over medium heat, remove chicken.
4. Reheat oil, deep fry chicken again over high heat for 10 seconds. Remove chicken and drain off the oil. Place on a plate.
5. Pour oil away, use only 1 tablespoon of oil to boil cranberry juice together. Pour sauce over chicken, serve hot.

番茄起司烤鸡胸
Baked Chicken with Tomato Sauce

材料：
鸡胸肉2片、比萨起司3大匙

番茄酱汁：
番茄1个、洋葱末2大匙、番茄膏2茶匙、番茄酱2大匙、意大利香料 1/2 茶匙、盐和胡椒适量调味

调味料：
（1）溶化的奶油（或橄榄油） 1/4 杯、大蒜泥1大匙、盐 1/3 茶匙
（2）面包粉 1/2 杯、胡椒粉 1/4 茶匙、Parmesan 起司粉2大匙

做法：
1. 番茄去皮、去籽，切成小丁，再加入其他的番茄酱汁材料拌匀。
2. 另外在2个碗中，分别将调味料（1）和（2）调匀备用。
3. 鸡胸肉先在牛油料中浸5分钟，再蘸裹上面包粉料。
4. 把鸡胸肉排放在烤盘中（烤盘中涂少许油），涂上番茄酱汁，盖上一张铝箔纸。放入预热至220℃的烤箱中，烤30分钟。
5. 取出烤盘，撒上比萨起司，再烤10～15分钟至起司溶化。
6. 烤鸡胸盛装在餐盘上，附炒蔬菜上桌。

Ingredients:
2 pieces boneless & skinless chicken breast, 3 tbsp pizza cheese

Tomato Sauce:
1 tomato, 2 tbsp chopped onion,
2 tsp tomato paste, 2 tbsp ketchup, 1/2 tsp Italian seasoning, salt & pepper to taste

Seasonings:
(1) 1/4 cup melted butter (or olive oil), 1/3 tsp salt,
   1 tbsp mashed garlic
(2) 1/2 cup bread crumbs, 1/4 tsp black pepper,
   2 tbsp parmesan cheese

Procedures:
1. Peeled tomato, seeded and then chop it. Mix with other ingredients of tomato sauce.
2. In two bowls, combine seasonings (1) and (2) separately.
3. Dip chicken in butter mixture for 5 minutes, then coat with bread crumb mixture.
4. Place chicken on a pan (butter it first), apply tomato sauce on top of chicken, cover with foil. Preheat oven to 220℃, bake chicken for 30 minutes.
5. Remove foil, top with pizza cheese. Bake for 10～15 minutes until cheese melt. Remove.
6. Serve with stir-fried vegetables.

Caesar Salad

Caesar salad dressings:
1 egg yolk, 4~5 capers, 1/2 tsp mashed garlic, 1 tsp yellow French mustard, 2/3 cup olive oil,
1 tbsp lemon juice, 1/2 tsp Tabasco sauce, salt & black pepper to taste
Combine egg yolk, capers, mashed garlic, and mustard in a bowl, add olive oil in, stir slowly and constantly to form a pasty and creamy salad dressing. Add lemon juice, Tabasco sauce, salt, pepper and Parmesan cheese to taste.

凯萨鸡肉沙拉
Chicken Caesar Salad

材料：
鸡胸肉1片、生菜200克、小番茄8个、培根2片、
吐司面包1片、Parmesan 起司粉1大匙

调味料：
（1）盐 $1/2$ 茶匙、胡椒粉 $1/2$ 茶匙、酒2大匙、水 $1/2$ 杯
（2）凯萨沙拉酱4～5大匙

做法：
1. 鸡胸肉冲洗一下，沥干水分，用调味料（1）浸泡，放置20分钟。
2. 烤箱预热至220℃，放入鸡胸肉烤12～15分钟至熟，依鸡胸肉的厚薄，把烤的时间略作增减。也可以用微波炉微波至熟。取出切片。
3. 生菜泡冰水冰镇10分钟，切成段；小番茄切半；吐司面包切成小丁，在预热烤箱时即可放入，烤成金黄色，取出放凉。
4. 培根切丝，用少许油煎至脆，用纸巾吸干油分。
5. 将生菜、番茄和鸡胸肉放入碗中，加入凯萨沙拉酱，拌匀后装盘，再撒上培根碎、面包丁和起司粉。

凯萨沙拉酱：
材料：蛋黄1个、酸豆4～5粒、大蒜泥 $1/2$ 茶匙、黄色芥末1茶匙、橄榄油约 $2/3$ 杯、柠檬汁1大匙、Tabasco辣椒水 $1/2$ 茶匙、盐和黑胡椒粉各适量。
做法：大碗中先将蛋黄、酸豆、大蒜泥、芥末酱搅拌均匀，慢慢地加入橄榄油，边加边搅拌，搅拌成酱汁。加入柠檬汁、辣椒水、盐、黑胡椒粉和起司粉调味即成为凯萨沙拉酱。

Ingredients:
1 piece chicken breast, 200g. romaine lettuce, 8 cherry tomatoes,
2 slices bacon, 1 slice bread, 1 tbsp Parmesan cheese

Seasonings:
(1) $1/2$ tsp salt, $1/2$ tsp pepper, 2 tbsp wine, $1/2$ cup water
(2) 4 ~ 5 tbsp Caesar salad dressing

Procedures:
1. Rinse and drain dry the chicken breast, brine with seasonings (1) for 20 minutes.
2. Preheat oven to 220℃, bake chicken for 12 ~ 15 minutes until done. The baking time might different according to the thickness of chicken breast. You may microwave chicken on high until done. Slice it after cools.
3. Rinse and soak romaine in ice water for 10 minutes. Cut to section. Halve tomatoes. Dice bread, bake until golden brown. Remove and let it cool.
4. Slice bacon, fry with a little of oil until crispy, remove.
5. Mix romaine, tomato, and chicken with Caesar dressing, place on a plate, top with bacon and toasted bread. Add more Parmesan cheese as desired.

Fried Chicken with Salsa and Tar-tar Sauce

酥炸鸡条佐双酱
Fried Chicken with Salsa and Tar-tar Sauce

材料：
鸡胸肉2片

调味料：
（1）盐 1/2 茶匙、胡椒粉 1/4 茶匙、蒜泥1茶匙、水 1/2 杯
（2）炸鸡粉 2/3 杯、脆酥粉 2/3 杯、胡椒粉 1/4 茶匙、水1杯

沙沙酱：番茄1个、洋葱末2大匙、香菜2根、红辣椒1个、柠檬汁1茶匙、盐和糖各少许适量调味

塔塔酱：美奶滋3大匙、法式芥末酱 1/2 茶匙、剁碎的酸豆1大匙、剁碎的酸黄瓜1大匙、剁碎的洋葱末2茶匙、白煮蛋 1/2 个（剁碎）

做法：
1. 将鸡胸肉较厚的地方再片开，用刀背拍打，使肉质松嫩，再切成条。
2. 把调味料（1）先放在盆中调匀，再放下鸡肉条腌15～20分钟。
3. 调味料（2）混合均匀，糊料太干时可以再加一点水，但要尽量浓稠。
4. 洋葱末泡水3分钟以去除辣气；番茄、香菜和红辣椒分别剁碎（或和洋葱一起放在调理机中，快速打几下），全部放入碗中，加盐、糖和柠檬汁调味，做成沙沙酱。
5. 另外调好塔塔酱。
6. 把4杯炸油烧至7分热，鸡肉条沥干水分并擦干后，蘸裹上粉糊料，投入油中，先以中火炸熟，捞出。
7. 油再烧至8～9分热，加入鸡条，以大火炸至酥脆且呈金黄色，捞出，沥干油渍，装盘。附沙沙酱和塔塔酱上桌。

Ingredients:
2 pieces chicken breast

Seasonings:
(1) 1/2 tsp salt, 1/4 tsp pepper, 1 tsp mashed garlic, 1/2 cup water
(2) 2/3 cup fry chicken powder, 2/3 cup crispy powder, 1/4 tsp pepper, 1 cup water

Salsa: 1 tomato, 2 tbsp chopped onion, 2 stalks cilantro, 1 red chili, 1 tsp lemon juice, sugar & salt to taste

Tartar Sauce: 3 tbsp. mayonnaise, 1/2 tsp. yellow French mustard, 1 tbsp. chopped capers, 1 tbsp. chopped pickles, 2 tsp. chopped onion, 1/2 boiled egg (chopped)

Procedures:
1. Slice the thick part of chicken meat to make it into a thinner layer, then cut into strips.
2. Mix seasonings (1) with chicken for 15 ~ 20 minutes.
3. Mix seasonings (2) in a bowl, add more water if it is too dry, make it into a thick paste.
4. Soak chopped onion for 3 minutes, drain. Chop tomato, cilantro and red chili, mix together (or blend them roughly with onion). Season with salt, sugar and lemon juice. This is the salsa.
5. Make Tar-tar sauce in another bowl.
6. Heat 4 cups of oil to 140℃. Coat chicken with paste (drain dry the chicken first), deep-fry chicken over medium heat until done. Remove.
7. Reheat oil to 160 ~ 170℃, deep fry chicken again over high heat until crispy. Drain and then arrange on a plate. Serve with salsa and Tar-tar sauce.

Stir-fried Minced Chicken

生炒鸡松
Stir-fried Minced Chicken

材料：
鸡肉250克、香菇3朵、笋丁1杯、韭黄丁1杯、青豆 1/2 杯、鸡蛋2个、油条1根或米粉1小片（约40克）、西生菜1球

调味料：
（1）盐 1/4 茶匙、淀粉1茶匙、水2大匙
（2）盐 1/3 茶匙、淡色酱油1大匙、淀粉 1/2 茶匙、水3大匙、胡椒粉少许、麻油 1/2 茶匙

做法：

1. 将油条切片，放入烤箱中烤至酥脆，取出放凉；如果用米粉，就要放入烧得极热的油中炸至泡起（每面各炸3秒钟）。捞出后沥干油，放在盘中，略为压碎。
2. 鸡肉切成小丁，用调味料（1）拌好，放置5~10分钟。
3. 香菇泡软、切成小丁；鸡蛋打散，煎成蛋皮后切成小丁；西生菜（生菜的一种）修剪成圆形小碗状。
4. 锅中把1杯油烧至7分热，放下鸡丁炒熟后盛出。油倒出，仅留下约2大匙。
5. 放下香菇丁，小火炒香，再放入笋丁和青豆同炒。加入鸡丁与蛋皮丁，淋下调味料（2），改大火拌炒均匀。
6. 最后将韭黄丁撒下即熄火，再略加拌和即可盛出，放在油条上，附生菜叶上桌包食。

Ingredients:
250g. chicken breast, 3 dried shitake mushrooms, 1 cup diced bamboo shoot, 1 cup diced white leeks, 1/2 cup green peas, 2 eggs, 1 piece yu-tiao or 40g. rice noodles, 1 head lettuce,

Seasonings:
(1) 1/4 tsp salt, 1 tsp cornstarch, 2 tbsp water
(2) 1/3 tsp salt, 1 tbsp light colored soy sauce, 1/2 tsp cornstarch, 3 tbsp water, pinch of pepper, 1/2 tsp sesame oil

Procedures:

1. Cut yu-tiao to small pieces, bake until crispy, remove and let cool. For rice noodle, deep fry it in very hot oil until puffed and light browned (only 2 or 3 seconds for each side). Remove and put on a platter. Crush them with a fork or chopsticks.
2. Cut the chicken into small cubes. Marinate with seasonings (1) for about 5~10 minutes.
3. Soak shitake mushrooms to soft, then dice it. Make a thin pancake with the beaten egg, cut into cubes. Trim lettuce to round shape as a small bowl.
4. Heat 1 cup of oil to 140℃, stir-fry chicken until done, remove chicken and pour the oil away, keep only 2 tablespoons of oil in wok.
5. Add mushroom, sauté until fragrant, add bamboo shoot and green peas, stir-fry for a while. Add chicken and egg in, stir-fry until mixed. Pour seasonings (2) in, stir-fry over high heat.
6. Add leek in and turn off the heat, mix evenly with all ingredients. Remove and place on top of yu-tiao, serve with lettuce.

杂菜鸡丁
Stir-fried Chicken with Vegetables

材料：
鸡胸肉200克、洋菇8朵、榨菜丁3大匙、熟笋丁3大匙、
大蒜末1茶匙、芹菜2根、葱花1大匙、热狗面包或蛋饼或烧饼3～4个

调味料：
（1）盐 1/4 茶匙、胡椒粉少许、水1大匙、淀粉1茶匙
（2）酱油1大匙、酒1大匙、糖 1/3 茶匙、水3大匙、胡椒粉少许、麻油 1/4 茶匙

做法：
1. 可选购绞好的鸡胸肉比较方便，或是自己切成小丁，拌上调味料（1）。
2. 洋菇切粒；芹菜也切粒。
3. 锅中先用2大匙油将鸡肉炒散，盛出。加少许油，再放入大蒜、葱花和洋菇炒香。
4. 加入榨菜和笋丁再炒一下，放回鸡肉，并依次加入酱油、酒、糖和水炒匀，最后加入芹菜粒，再加入胡椒粉和麻油，拌炒均匀。
5. 热狗面包或烧饼烤热，或把蛋饼皮煎熟，和杂菜鸡丁一起上桌包食。

Ingredients:
200g. chicken meat, 8 pieces mushroom, 3 tbsp chopped preserved mustard, 3 tbsp diced bamboo shoot (cooked), 1 tsp chopped garlic, 2 stalks celery, 1 tbsp chopped green onion, 4 hot dog rolls or baked buns or egg roll wrappers

Seasonings:
(1) 1/4 tsp salt, pinch of pepper, 1 tbsp water, 1 tsp cornstarch
(2) 1 tbsp soy sauce, 1 tbsp wine, 1/3 tsp sugar, 3 tbsp water, a pinch of pepper, 1/4 tsp sesame oil

Procedures:
1. You may buy the ground chicken meat or just cut it by yourself. Mix with seasonings (1).
2. Dice mushrooms; cut celery to small cubes.
3. Heat 2 tablespoons of oil to stir-fry chicken, remove when it is cooked. Add garlic, green onion, and mushroom in, sauté until fragrant.
4. Add preserved mustard and bamboo shoot in, stir for a while. Return chicken, add soy sauce, wine, sugar, and water in, stir-fry evenly over high head. Add celery, pepper and sesame oil at last, mix well. Remove.
5. Bake hot dog rolls or buns, or fry the egg roll wrappers with a little of oil until done, serve together with the chicken.

Minced Chicken with Abalone Potage

鸡茸鲍鱼羹
Minced Chicken with Abalone Potage

材料：
罐头鲍鱼 1/2 罐、鸡里脊肉 4~5 条（约150克）、蛋白4个、熟火腿末1大匙、面粉4大匙、清汤5杯

调味料：
酒 1/2 大匙、盐1茶匙

做法：
1. 将鸡里脊肉用刀刮成细茸状(见小图)，再用刀背剁细一点，放入大碗内。
2. 加入酒及 1/3 茶匙的盐，然后加入1个蛋白搅拌，需朝同一个方向轻轻搅拌，待蛋白完全被鸡茸吸收后，再加入第二个蛋白，再加以搅拌、吸收，依次将 4 个蛋白完全拌入鸡茸中。
3. 鲍鱼切成大薄片，罐中的鲍鱼汤汁取用约1杯。
4. 锅中烧热3大匙油，以小火炒香面粉，炒至匀滑后，慢慢地倒下清汤及鲍鱼汤汁，边倒边搅，调拌均匀，不要有小颗粒。
5. 先用大火煮滚，放下鲍鱼片，加盐调味，待再煮滚时，将鸡茸慢慢倒入汤中，一边倒，一边搅动汤汁，倒完之后即关火，以免鸡茸变老。
6. 盛装至大汤碗内或深底菜盘中，再撒下熟火腿末即可。

Ingredients:
1/2 can abalone, 4 ~ 5 pieces chicken tender (about 150g.), 4 eggs (only use egg white), 1 tbsp chopped Chinese ham (cooked), 4 tbsp flour, 5 cups chicken stock

Seasonings:
1/2 tbsp wine, 1 tsp salt

Procedures:
1. Scrape chicken tender with knife (see picture), chop with the back edge of knife to very fine, place in a large bowl.
2. Add wine, 1/3 teaspoon of salt and 1 egg white in, stir in one direction to mix evenly with the chicken meat, then add the second egg white, stir until evenly mixed. Add one at a time, mix well between each addition.
3. Slice abalone into thin slices. Reserve 1 cup of the abalone broth from the can.
4. Heat 3 tablespoons of oil to stir fry flour over low heat for a few seconds until flour become light brown. Add in soup stock and the reserved abalone broth slowly.
5. Mix thoroughly and bring to a boil. Add abalone slices and season with salt. When soup boils again, pour in the minced chicken paste slowly. Stir the potage until thoroughly mixed. Turn off the heat immediately after pouring.
6. Pour the abalone potage into a soup bowl, sprinkle ham on top. Serve.

鸡茸烩瓜丝
Minced Chicken with Wax Gourd

材料：
冬瓜450克、鸡里脊肉3条（约120克）、
蛋白3个、清汤4杯、葱1根、姜2片、芹菜末或香菜末适量

调味料：
酒1大匙、盐1茶匙、淀粉水1大匙、白胡椒粉少许、麻油少许

做法：
1. 冬瓜去皮、去籽及较软绵的白肉，切成细丝。
2. 鸡里脊肉依照第96页的方法加蛋白搅成鸡茸。
3. 用1大匙油煎香葱段和姜片，放下冬瓜丝略炒，淋下酒，倒入清汤，烧至冬瓜透明，加盐调味，捡弃葱姜。
4. 用淀粉水勾薄芡，再把鸡茸淋入锅中，边淋边搅动，淋完后立即关火。撒下胡椒粉、滴下麻油，轻轻搅动一下。
5. 装碗后撒下芹菜末。

Ingredients:
450g. wax gourd, 3 pieces chicken tenderloin (about 120g.),
3 eggs (only use egg white),
4 cups chicken stock,
1 stalk green onion,
2 slices ginger,
chopped celery or cilantro

Seasonings:
1 tbsp wine, 1 tsp salt, 1 tbsp cornstarch paste,
a little of white pepper and sesame oil to taste

Procedures:
1. Peel and remove seeds from wax gourd, also remove some of the soft flesh from the inside. Shred it.
2. Mash chicken, mix with egg white to make chicken paste as the recipe on P. 99.
3. Heat 1 tablespoon of oil to fry green onion sections and ginger, add gourd in, stir-fry for a while. Pour wine and chicken stock, cook until gourd is soft. Season with salt, discard green onion and ginger.
4. Thicken with cornstarch paste. Pour chicken paste in, stir the soup while pouring. Turn off the heat right away after pouring. Add pepper and sesame oil.
5. Remove to a soup bowl, add celery or cilantro at last.

碎米鸡丁
Diced Chicken with Peanuts

材料：
鸡胸肉1片（约250克）、四川泡菜1杯、
花生米 1/4 杯、泡红椒或新鲜红辣椒2个、葱花1大匙

调味料：
（1）盐 1/3 茶匙、水1～2大匙、淀粉1茶匙
（2）淡色酱油1茶匙、酒 1/2 大匙、盐 1/4 茶匙、糖 1/4 茶匙、醋 1/2 茶匙、水3大匙、淀粉 1/4 茶匙

做法：
1. 鸡胸肉去皮和筋后切成小丁，用调味料（1）拌匀，腌20分钟。
2. 泡菜剁成小丁；泡辣椒切丁；花生米去皮后可以略加剁碎。
3. 锅中将1杯油烧至5～6分热，放入鸡丁泡炒至熟，捞出、沥净油。油倒出，仅留下1大匙左右。
4. 放入泡辣椒和葱花炒香，加入泡菜和鸡丁一起炒热，再加入调匀的调味料（2），大火快炒均匀，关火，撒下花生米即可装盘。

Ingredients:
250g. chicken breast, 1 cup Sichuan pickles, 1/2 cup peanuts, 2 pickled red chilies or fresh red chilies, 1 tbsp chopped green onion

Seasonings:
(1) 1/3 tsp salt, 1～2 tbsp water, 1 tsp cornstarch
(2) 1 tsp light colored soy sauce,
　　1/2 tbsp wine, 1/4 tsp salt,
　　1/4 tsp sugar, 1/4 tsp vinegar,
　　3 tbsp water, 1/4 tsp cornstarch

Procedures:
1. Trim off skin from chicken meat then dice it, marinate with seasonings (1) for 20 minutes.
2. Chop pickles and pickled red chilies to small pieces; you may crush peanuts slightly or keep them in whole shape.
3. Heat 1 cup of oil to 100～120℃, stir-fry chicken until done, drain. Pour away the oil, just keep 1 tablespoon of oil in wok.
4. Add in green onion and pickled red chili, stir-fry until fragrant. Add pickles, chicken and seasonings (2), stir-fry over high heat until well mixed. Turn off the heat. Sprinkle peanuts over chicken, mix slightly and serve.

泰式辣炒鸡肉
Spicy Chicken, Thai Style

材料：
鸡胸肉250克、小番茄6颗、蒜末 1/2 茶匙、大辣椒 1/2 个、九层塔少许

泰式辣椒酱：
红辣椒末2大匙、洋葱末1大匙、蒜末1茶匙

调味料：
蚝油1茶匙、鱼露2茶匙、糖 1/2 茶匙、胡椒粉少许、酒1茶匙、高汤或水2大匙

做法：
1. 用2大匙温油炒辣椒末、洋葱末和大蒜末，以中火慢慢炒香，盛出备用。
2. 鸡肉剁碎；小番茄切成两半；大辣椒去籽、切片。
3. 锅中放约1大匙的油，加入小番茄、蒜末、大辣椒片和炒好的辣椒酱，用小火再炒香。
4. 加入所有的调味料和剁碎的鸡胸肉，用大火炒煮2分钟，至鸡肉已全熟。
5. 起锅时，加入九层塔叶，略为拌和即可装盘。

Ingredients:
250g. chicken breast, 6 cherry tomatoes,
1/2 tsp chopped garlic, 1/2 red pepper, basil leaves

Thai style chili paste:
2 tbsp chopped red chili,
1 tbsp chopped onion,
1 tsp chopped garlic

Seasonings:
1 tsp oyster sauce, 2 tsp fish sauce,
1/2 tsp sugar, a pinch of pepper,
1 tsp wine, 2 tbsp soup stock or water

Procedures:
1. Stir-fry chopped red chili, onion and garlic with 2 tablespoons of warm oil over medium low heat until fragrant, remove to a bowl; this is the Thai style chili paste.
2. Chop chicken to small pieces; halve tomato; remove seeds from red pepper, then slice it.
3. Heat 1 tablespoon of oil to stir-fry tomatoes, chopped garlic, red pepper and Thai style chili paste over low heat until fragrant.
4. Add all seasonings and chicken meat, stir-fry over high heat for about 2 minutes until chicken is cooked.
5. Turn off the heat, add basil leaves at last, mix slightly and serve.

咸鱼鸡粒炒饭
Stir-fry Rice with Salty Salmon & Chicken

材料：
鸡胸肉80克、咸鲑鱼100克、西生菜2片、白饭2碗、葱花1大匙

调味料：
（1）盐少许、水 1/2 大匙、淀粉 1/4 茶匙
（2）盐 1/2 茶匙、白胡椒适量

做法：
1. 鸡胸肉切小粒，用调味料（1）拌匀，腌15～20分钟。
2. 咸鲑鱼切成小丁，用油炒炸至酥，盛出备用；西生菜切成小丁。
3. 鸡胸肉用2大匙油炒熟，盛出。放入白饭，以中小火慢慢把饭炒热并炒散开，加盐调味。
4. 放入鸡丁、咸鲑鱼和葱花，再炒拌一下后关火。放入西生菜丁，并撒下白胡椒粉，再拌炒均匀即可。

※把约1茶匙的盐抹在一片鲑鱼上，放入塑胶袋中包好，上面压一个重的东西，腌1天后即是咸鲑鱼。也可以用其他的咸鱼来炒饭。

Ingredients:
80g. chicken breast, 100g. salty salmon,
2 pieces lettuce leaf, 2 cups cooked rice,
1 tbsp chopped green onion

Seasonings:
(1) pinch of salt, 1/2 tbsp water,
    1/4 tsp cornstarch
(2) 1/2 tsp salt, pinch of white pepper

Procedures:
1. Cut chicken to small pieces, marinate with seasonings (1) for 15～20 minutes.
2. Dice salty salmon to small cubes, fry with hot oil until crispy, remove. Cut lettuce into small pieces.
3. Stir-fry chicken with 2 tablespoons of oil, remove when it is cooked. Add rice, stir-fry over medium-low heat until rice is heated up and all separate. Season with salt.
4. Put chicken, salty salmon and green onion in, stir-fry evenly. Turn off the heat, add lettuce and pepper, stir-fry evenly.

※Marinate 1 piece of salmon with 1 teaspoon of salt for one day to make salty salmon. Or you may use other kind of salty fish.

细说鸡腿
All about Chicken legs

　　鸡腿是中国人比较喜爱的部分，因为常活动的关系，所以这部分肉有弹性、嫩又多汁，适合用的烹调方法非常广泛，因此我把它分为去骨鸡腿和带骨鸡腿两个篇章来介绍。去骨鸡腿可以用来炒、煎、炸或蒸，如果因为个人因素喜爱鸡胸肉的话，许多菜式也可以用鸡胸肉来代替鸡腿肉。

Chinese prefer chicken legs. Due to the exercise, chicken legs are tender and juicy. They are suitable for many different cooking methods, and therefore I separate this section into two parts: "boneless chicken legs" and "bone-in chicken legs". In many dishes, you may substitute chicken legs with breasts if you prefer.

Part I 去骨鸡腿 Boneless Chicken Legs

鸡腿去骨的方法：

　　沿着腿骨的左右两边深深切划开来，使腿骨露出来（图1）；翻过鸡腿，把前段骨头剔出（图2）；接着在关节骨处敲一刀，敲断关节骨（图3），再翻面露出腿骨，一手拿刀子压住腿骨，一手拉住关节骨，用力一拉（图4），便可使肉与骨头分离，再把腿骨剔下来（图5）。去骨技巧需要练习，当然也可以请鸡贩代为处理，或在超市选购已经去好骨的腿肉（图6），但是事实上并不难，读者不妨试一试。

How to de-bone:

Slice the chicken leg open along the bone and expose the bone (pic 1). Flip the leg over and separate the meat from the thigh bone (pic 2). Break the joint with knife (pic 3) and pull out the thigh bone (pic 4). Separate the bone from the rest of the meat with a knife (pic 5). Debone takes practice; however it is not as difficult as it seems. Of course you can always ask your butcher to do the work for you, or choose to purchase boneless chicken legs that have been prepared in supermarkets（pic 6）.

图1

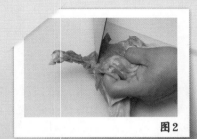
图2

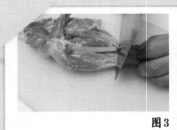
图3

图4

图5

图6

去骨鸡腿的处理：

去骨鸡腿多半用来炒鸡丁，或者整只烹调做成鸡腿排。因为已经没有腿骨的支撑，所以腿部的筋络遇热就会收缩，使得鸡腿排缩短，既不好看肉质也比较老，因此在剔除大骨之后，最重要的就是先在带白筋的部分（鸡的棒棒腿部分），以和白筋成垂直的方向斩剁几刀，把筋剁断，使它不会再收缩；也把肉质松弛一下，再来烹调。同时在鸡腿的上半部肉比较厚的地方也要剁几下，有了刀口，肉才容易入味，也熟得快（图7）。剁好刀口后就可以剁成条或块来烹调。

烹调整只去骨鸡腿排时，有一点很重要的就是要保留鸡腿的关节骨（图8），利用关节骨牵引住部分腿部的筋，使鸡腿肉不至于收缩得太短才好。做鸡腿排的菜式一般是整只烹煮好、定型之后再做分割。

How to process:
De-boned chicken legs are commonly cooked whole or chopped into cubes for stir-frying. Without the support of the bone, chicken legs shrink notably when they are cooked which toughens the meat and affects the appearance. Therefore it is important to make a few cuts perpendicular to the leg muscles to prevent shrinkage. This also relaxes the meat and allows it to absorb the marinade better and cook faster (pic 7). To cube or cut to pieces before marinate it.

When cooking boneless chicken legs whole, it is important to leave the ankle joint of the drumstick on (pic 8). It is common to cook the boneless chicken leg in whole piece to keep the shape, and then slice it to the desired pieces.

图7

图8

图9

鸡腿肉的腌泡方法：

　　去骨鸡腿多半用来炒或煎、炸，因此常选用白肉鸡和仿土鸡（半土鸡）。腿肉在炒熟后属于深色，因此可以用酱油来腌，不但有咸味，同时还带有香气（图9）。

　　腌的时间长短和鸡块大小有关，如果加了淀粉一起腌，最好放置30分钟以上，以使淀粉能稳定地附着在鸡肉上。冰箱温度适宜时，腌过的鸡肉可以保存3天。

How to marinate:
Boneless chicken legs are commonly for stir-fried, sautéed, or deep-fried. Since chicken legs are dark meat, you may marinate them with soy sauce, which provides saltiness as well as aroma.

The amount of time to marinate depends on the size of the meat. If cornstarch is added as part of the marinade, the meat should be marinated for at least 30 minutes so that the cornstarch can settle on the meat. Marinated chicken may be kept for 3 days in refrigerator.

烹调重点：

炒之前仍然需要先过油以使肉质香嫩，油温大多是以8～9分热（160～180℃）为准，油温通常是依照鸡肉量的多少、用油的量和火力的大小而做调整。过油至8分熟后捞出鸡肉，再去炒制。当然，和鸡胸肉一样，你也可以改用滚水汆烫至8分熟，虽然香气不如过油的鸡肉，但是热量却大为减少。

Tips for cooking:

Pre-stir-frying is recommended for achieving tenderness and more aroma. According to the quantity of meat, the amount of oil and heat power vary, the temperature of oil is typically 160–180 ℃. In other way, you may blanch the meat to half cooked as I mentioned in chicken breast part, although it will be less in fragrant but also less in calorie.

Wined Chicken with Chinese Herbs

去骨鸡腿

参杞醉鸡卷
Wined Chicken with Chinese Herbs

材料：
仿土鸡腿2只、枸杞子1大匙、黄芪5～6片、铝箔纸2张
调味料：
鱼露或虾油6大匙、绍兴酒1杯
做法：
1. 鸡腿剔除大骨，将肉较厚的地方片开，使肉薄一些。放入盆中，加入3大匙鱼露和1/4杯水，腌泡1～2小时。
2. 把鸡腿卷成长条，用铝箔纸包卷好，两端扭紧，好像一颗糖果。
3. 蒸锅中水煮滚，把鸡腿卷放入蒸锅内，以中火蒸45～50分钟，取出。
4. 在一个深盘中，放下3大匙鱼露、绍兴酒和冷开水1杯调匀，加入枸杞子和黄芪。
5. 待鸡腿卷完全凉透后，取出鸡腿，连汤汁一起加入酒中浸泡（见小图）。
6. 用保鲜膜密封盘子或盖上盖子，放入冰箱中冷藏，2～3小时后便可食用，最好浸泡1天使鸡腿更入味。可保存3～4天。
7. 吃时取出鸡腿，切片排盘，淋上一些酒汁。

Ingredients：
2 simulate native chicken legs, 1 tbsp medlar (gou-qi-zi),
5～6 slices huang-qi, 2 pieces foil
Seasonings：
6 tbsp fish paste or shrimp paste, 1 cup shaoxing wine
Procedures：
1. Remove the bones of the chicken leg (keep the join in drum stick), trim off the thick part of meat, make the thickness evenly as possible as you can. Marinate with 3 tablespoons of fish paste and 1/4 cup of water for 1～2 hours.
2. Roll the chicken to a cylinder, pack with a piece of foil tightly. Make two rolls.
3. Steam chicken rolls for 45～50 minutes over medium heat. Remove chicken.
4. Combine 3 tablespoons of fish paste with wine and 1 cup of water in a deep plate to make wine broth, add medlar and huang-qi in.
5. Unpack the rolls when cool, soak chicken legs in wine broth (add the chicken broth to wine broth, see as picture).
6. Seal or cover the plate. Soak for 2～3 hours at least, it will be more tasteful after soaking for one day. It may keep for 3～4 days.
7. Cut to slices before serving, arrange on a plate, pour some wine broth over chicken.

Diced Chicken with Cashew

去骨鸡腿
107

腰果鸡丁
Diced Chicken with Cashew

材料：
去骨鸡腿2只、生腰果 2/3 杯、青椒 1/2 个、
红甜椒 1/2 个、葱2根、姜6～8小片

调味料：
（1）酱油1大匙、水2大匙、淀粉 1/2 大匙
（2）酱油1大匙、酒 1/2 大匙、糖 1/2 茶匙、醋 1/4 茶匙、水3大匙、麻油数滴

做法：
1. 在鸡腿的表面上轻轻剁过，再切成2厘米的鸡丁，用调味料（1）拌匀，腌30分钟左右。
2. 青椒及红甜椒去籽，切成块；葱切成段。
3. 腰果用糖水浸泡1小时，捞出，用5分热的油，小火慢慢炸熟，炸3～4分钟，捞出。也可以把腰果放入烤箱中，以120℃烤至熟，取出、放凉。
4. 锅烧热后，放入1杯油，待油烧至8分热时，将鸡肉倒入锅中，大火过油，炸约1分钟，至鸡肉转为浅色，捞出。油倒出，仅留下1大匙。
5. 用大火先爆炒葱段、姜片及青椒、红甜椒，倒下鸡丁同炒数下，淋下调味料（2），大火拌炒均匀，关火，加入腰果一拌即可装盘。

※也可以买烤好的熟腰果，或是核桃、夏威夷果等其他核果类。

Ingredients:
2 chicken legs (de-boned), 2/3 cup cashew, 1/2 green pepper,
1/2 red bell pepper, 2 stalks green onion, 6 ~ 8 slices ginger

Seasonings:
(1) 1 tbsp soy sauce, 2 tbsp water, 1/2 tbsp cornstarch
(2) 1 tbsp soy sauce, 1/2 tbsp wine, 1/2 tsp sugar, 1/4 tsp vinegar, 3 tbsp water, few drops of sesame oil

Procedures:
1. Cut chicken into 2 cm pieces. Marinate with seasonings (1) for at least 30 minutes.
2. Remove seeds from green and red pepper, then cut to pieces. Cut green onion to sections.
3. Soak cashew in light syrup for 1 hour. Deep-fry in 100℃ oil for about 3 ~ 4 minutes or bake it over 120℃ until done. Set aside and let cool.
4. Heat 1 cup of oil to 160℃ to deep fry chicken. Fry it over high heat for 1 minute. Remove chicken and drain off oil.
5. Heat 1 tablespoon of oil to stir fry ginger slices and green onion, when fragrant, add green and red pepper, stir-fry for a while. Add chicken and seasonings (2), stir-fry over high heat until thoroughly mixed. Turn off the heat and add cashew at last. Serve hot.

※You may use the roasted cashew or walnut or Hawaii macadamia nut or other kinds of nut you like.

去骨鸡腿

Diced Chicken with Basil & Celery

去骨鸡腿

香芹九层鸡丁
Diced Chicken with Basil & Celery

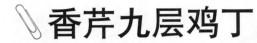

材料：
去骨鸡腿2只、西芹2棵、姜片15片、大蒜6粒、红辣椒1个、九层塔4～5枝

调味料：
（1）酱油2大匙、糖 1/2 茶匙、水2大匙、淀粉1大匙
（2）酒1大匙、酱油膏2大匙、糖 1/2 茶匙、水4大匙

做法：
1. 鸡腿肉切成2～3厘米大小的块，用调味料（1）拌匀，腌30分钟。
2. 西芹削去老筋、斜切成片；大蒜切片；红辣椒切斜段；九层塔摘嫩叶。
3. 锅中烧滚4杯水，放下鸡丁，用大火烫至8分熟，捞出，放下西芹烫5秒钟，捞出。
4. 锅中放2大匙油，加入姜片与大蒜，慢火煎炒至香气透出，倒回鸡丁、西芹和红椒段再炒，淋下调味料（2），焖煮半分钟至鸡熟。
5. 撒下九层塔、关火，再拌匀便可装盘。

Ingredients：
2 chicken legs (de-boned), 2 sticks celery, 15 ginger slices,
6 cloves garlic, 1 red chili, 4 ~ 5 stalks basil leaf

Seasonings：
(1) 2 tbsp soy sauce, 1/2 tsp sugar, 2 tbsp water, 1 tbsp cornstarch
(2) 1 tbsp wine, 2 tbsp soy sauce paste, 1/2 tsp sugar, 4 tbsp water

Procedures：
1. Cut chicken to 2 ~ 3 cm pieces, marinate with seasonings (1) for 30 minutes.
2. Trim celery, cut to slices; slice garlic and red chili; trim basil leaves.
3. Bring 4 cups of water to a boil, blanch chicken over high heat until almost done, remove. Add celery in, blanch for only 5 seconds, drain.
4. Add 2 tablespoons of oil in a wok, fry ginger and garlic over medium-low heat until fragrant. Add chicken, celery and red chili, stir-fry for a while. Add seasonings (2), cook for 30 seconds until chicken is cooked.
5. Add basil leaves at last, turn off the heat and mix basil with chicken, remove to a plate.

去骨鸡腿

Chicken with Gung-Bao Sauce

※Gung-Bao Sauce is a very famous sauce in Sichuan, China. It is created by Mr. Ding Bao-Zhen in Qing Dynasty. The use of dry-fried chilies is a distinct characteristic of dishes with Gung-Bao Sauce.

去骨鸡腿

宫保鸡丁
Chicken with Gung-Bao Sauce

材料：

去骨鸡腿2只、油炸或烤熟的花生米 1/2 杯、干红辣椒10个、花椒粒1茶匙、姜末1茶匙

调味料：

（1）酱油1大匙、淀粉 1/2 大匙、水2大匙

（2）酱油1 1/2 大匙、深色酱油1茶匙、酒1大匙、糖2茶匙、醋1茶匙、淀粉 1/2 茶匙、水3大匙、麻油 1/4 茶匙

做法：

1. 在鸡腿的表面上用刀轻轻剁几下，把腿部的白筋剁断，同时使肉松弛后切成2.5厘米大小的块状，用调味料（1）拌匀，腌30分钟。
2. 干辣椒切成段；花生米去皮备用。
3. 锅中把1杯油烧至8～9分热，放下鸡肉过油炸熟，捞出。
4. 油倒出，只留下约1大匙油，先用小火炒香花椒粒，待成为深褐色时捞弃花椒粒。
5. 加入1大匙油，再放入辣椒段，小火炸至深红褐色，盛出。加入姜末和鸡肉，大火炒数下。
6. 加入调味料（2）炒匀，熄火后加入干辣椒和花生米，拌和后即可装盘。

※这是一道四川名菜，相传是在清朝时，官拜"宫保"的丁宝桢待客时所喜爱做的一道菜，因而流传下来，用干辣椒烹调是这道菜的特色。

Ingredients:

2 de-boned chicken legs, 1/2 cup peanuts (deep fried or roasted),
10 pieces dried red chili, 1 tsp brown peppercorn, 1 tsp chopped ginger

Seasonings:

(1) 1 tbsp soy sauce, 1/2 tbsp cornstarch, 2 tbsp water

(2) 1 1/2 tbsp soy sauce, 1 tsp dark color soy sauce, 1 tbsp wine, 2 tsp sugar, 1 tsp vinegar, 1/2 tsp cornstarch, 3 tbsp water, 1/4 tsp sesame oil

Procedure:

1. Chop lightly on the meat side of chicken with a knife to make chicken tender. Cut into 2.5 cm cubes, marinate with seasonings (1) for 30 minutes.
2. Wipe the dried red chilies, cut into sections; peel the peanuts.
3. Heat 1 cup of oil to 160 ~ 180℃ to fry chicken for about 1 minute. Remove chicken.
4. Keep only 1 tablespoon of oil to fry brown peppercorns over low heat until fragrant and dark brown, discard peppercorns.
5. Add 1 tablespoon of oil to fry dried chilies until more red, remove dried chilies. Add in ginger and chicken, stir fry over high heat quickly.
6. Add seasonings (2), stir until evenly mixed. Turn off the heat. Add dried chili and peanuts, mix well and serve.

左宗棠鸡
Stir-fried Chicken, Hunan Style

材料：
去骨鸡腿2只、红辣椒3个、大蒜5粒、姜末1茶匙
调味料：
（1）酱油 1/2 大匙、淀粉 1/2 大匙、水2大匙
（2）酱油1大匙、酒1大匙、淀粉 1/2 茶匙、水2大匙
（3）醋 1/2 大匙、麻油 1/3 茶匙
做法：
1. 鸡腿肉切成长条块，用调味料（1）拌匀，腌20～30分钟。
2. 红辣椒去籽，切成长条；大蒜切片。
3. 将锅先烧热，放入1杯油，待油达9分热时，将鸡肉倒入锅中，大火过油。待鸡肉变色有9分熟时，捞出。
4. 将油倒出，仅用1大匙油来爆炒大蒜片、姜末及红辣椒，至红辣椒微软时，倒下鸡肉同炒数下，淋下调味料（2），大火快速拌炒均匀。
5. 临起锅前，沿着锅边淋下醋烹香，再滴下麻油，炒匀即可。

※这道菜是正宗的湖南名菜，左公即清朝名将左宗棠将军，本菜的特点是鸡肉香、嫩。

Ingredients:
2 boneless chicken legs, 3 red chili peppers,
5 cloves garlic, 1 tsp chopped ginger
Seasonings:
(1) 1/2 tbsp soy sauce,
    1/2 tbsp cornstarch, 2 tbsp water
(2) 1 tbsp soy sauce, 1 tbsp wine,
    1/2 tsp cornstarch, 2 tbsp water
(3) 1/2 tbsp vinegar, 1/3 tsp sesame oil
Procedures:
1. Cut chicken meat into long pieces. Marinate with seasonings (1) for 20 ~ 30 minutes.
2. Remove the seeds from red chili peppers, cut into stripes. Slice the garlic cloves.
3. Heat the wok, then add 1 cup of oil in, heat to 180°C, fry chicken until almost done. Drain.
4. Heat 1 tablespoon of oil to sauté garlic, ginger and red pepper. Add chicken in, stir-fry for lots of seconds. Pour in the seasonings (2) and mix thoroughly.
5. Drizzle vinegar toward the edge of the wok to enhance the fragrant, add sesame oil at last. Mix evenly and then serve hot.

※This is a very famous Hunan dish which created by Mr. Zuo, zongtang in Qing Dynasty.

去骨鸡腿

香蒜烹鸡块
Fried Chicken with Garlic Sauce

材料：
鸡腿2只、大蒜6粒、圆白菜丝1杯

调味料：
（1）酱油2大匙、酒1大匙、葱1根、姜2片、水1大匙
（2）美极酱油 1/2 茶匙、水4大匙、黑胡椒粉少许

做法：
1. 用刀将鸡腿肉上的白筋斩断，放入调味料（1）中腌30分钟（葱姜要略拍碎）。
2. 大蒜切片；圆白菜丝洗净、沥干，排入盘中。
3. 锅中烧热3大匙油，鸡皮朝下，把鸡腿放入锅中，煎至金黄，翻面再煎至金黄。
4. 淋下 2/3 杯水，煮开后改小火，盖上锅盖，焖煮3分钟至鸡腿刚熟，取出鸡腿，改刀切成块。
5. 再加 1/3 杯水到锅中煮滚，加少许盐调味，淋在盘中的圆白菜丝上。
6. 在另一个锅中用1大匙油把大蒜炒香，放回鸡腿块一起再炒一下。
7. 滴下美极酱油和水烹煮一下，撒下胡椒粉，盛装到圆白菜上。

Ingredients:
2 boneless chicken legs, 6 garlic cloves,
1 cup cabbage shreds

Seasonings:
(1) 2 tbsp soy sauce, 1 tbsp wine,
    1 stalk green onion,
    2 slices ginger, 1 tbsp water
(2) 1/2 tsp Maggie sauce,
    4 tbsp water, pinch of black pepper

Procedures
1. Chop on the meat side of chicken to cut off the tendon, marinate with seasonings (1) for 1/2 hour (crush green onion and ginger first).
2. Slice garlic; rinse cabbage shreds, drain and then arrange on a plate.
3. Heat 3 tablespoons of oil in a pan, put chicken in with skin side down. Fry until the skin become golden brown. Turn it over, fry the meat side to golden brown too.
4. Add 2/3 cup of water in, turn to low heat after boiling. Cover the lid and simmer for 3 minutes until chicken are just cooked. Remove chicken, cut to pieces.
5. Add another 1/3 cup of water to the pan, season with salt, pour sauce over cabbage shreds.
6. Sauté garlic with 1 tablespoon of oil, return chicken to the pan, stir-fry with garlic.
7. Add in Maggie sauce and water, fry over high heat to enhance the fragrant. Sprinkle some black pepper at last. Remove to plate.

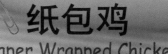

纸包鸡
Paper Wrapped Chicken

材料：
去骨鸡腿2只、熟火腿12片、
香菇2朵、香菜叶12枚、
玻璃纸12小张（10厘米正方）、麻油2大匙

调味料：
酱油2大匙、盐 1/4 茶匙、
糖 1/2 茶匙、酒1大匙、胡椒粉 1/4 茶匙

做法：
1. 鸡肉切成3厘米×5厘米的大薄片（共12片），全部放在碗里，加入调味料拌匀，腌15分钟。
2. 香菇用温水泡至软且透之后去菇蒂，每片切成6个三角形。
3. 熟火腿（可用西洋火腿）切成大小相仿的尖角小片。
4. 在玻璃纸的中央先刷上少许麻油，放上香菜叶、香菇及火腿各1片，然后盖上1片鸡肉。先将两个尖角对折，再将两边向内折起，包裹成长方形小包。
5. 将油烧至8分热，投下纸包鸡，需正面朝下投入油中。用小火慢炸，以免把玻璃纸炸焦，约炸2分钟。
6. 至鸡肉已变白而够熟时捞出，沥干油渍，排入盘内。

Ingredients:
2 boneless chicken legs, 12 slices ham,
2 dried shitake mushrooms,
12 parsley leaves,
12 pieces cellophane paper (10 cm × 10 cm),
2 tbsp sesame oil

Seasonings:
2 tbsp soy sauce, 1/4 tsp salt,
1/2 tsp sugar, 1 tbsp wine, 1/4 tsp black pepper

Procedures:
1. Cut the chicken into 12 slices, about 3 cm × 5 cm each. Marinate with seasonings for about 15 minutes.
2. Soak mushrooms to soft, then cut each into 6 small triangular pieces.
3. Slice the ham into pieces, the same size as mushroom.
4. Brush some sesame oil on cellophane paper, arrange one piece of parsley leaf, one piece of mushroom and one piece of ham on the paper, than place one piece of chicken on top, fold and pack it into a rectangular package.
5. Heat 4 cups of oil to 160℃ to deep-fry the chicken packages for about 2 minutes.
6. Remove chicken when cooked. Drain and place on a plate, serve.

去骨鸡腿

蚝油鸡丁蛋豆腐
Chicken and Egg Tofu in Oyster Sauce

材料：
去骨鸡腿1只、日本豆腐1盒、香菇2朵、葱1根

调味料：
（1）酱油 1/2 大匙、酒 1/2 大匙、淀粉1茶匙
（2）酒 1/2 大匙、蚝油 1/2 大匙、水 1/4 杯、麻油 1/4 茶匙

做法：
1. 鸡腿切成小块，用调味料（1）拌匀，腌20～30分钟。
2. 香菇泡软、切丁；葱切小段。
3. 日本豆腐切成方块。
4. 锅中用2大匙油把鸡肉炒熟，盛出，放下香菇和葱段炒香，淋下酒和蚝油再炒一下，加水煮滚。
5. 放下鸡丁和日本豆腐，轻轻搅拌，再煮一煮，如果汤汁仍多，可以勾点薄芡，滴下麻油即可。

Ingredients:
1 boneless chicken leg, 1 box egg tofu, 2 shitake mushrooms, 1 stalk green onion

Seasonings:
(1) 1/2 tbsp soy sauce, 1/2 tbsp wine, 1 tsp cornstarch.
(2) 1/2 tbsp wine, 1/2 tbsp oyster sauce, 1/4 cup water, 1/4 tsp sesame oil.

Procedures:
1. Cut chicken leg to small pieces, marinate with seasonings (1) for 20 ~ 30 minutes.
2. Soak shitake mushrooms to soft, then dice it. Cut green onion to small sections.
3. Cut tofu to pieces.
4. Heat 2 tablespoons of oil to stir-fry chicken, remove when cooked. Sauté mushrooms and green onion, add wine and oyster sauce, stir evenly. Add water, bring to a boil.
5. Add chicken and tofu in, mix gently, cook for a while. Thicken with cornstarch paste if needed. Drizzle sesame oil.

去骨鸡腿

沙茶鸡肉串
Bar-B-Q Chicken Skewers with Sha-cha Sauce

材料：
去骨鸡腿2只、竹签12支

调味料：
（1）酱油1大匙、酒1大匙、糖1茶匙、水2大匙
（2）花生酱1大匙、温水2大匙、沙茶酱2大匙、酱油1/2大匙

做法：
1. 鸡肉去皮后切成2厘米宽的长条，用调味料（1）拌匀，腌泡10分钟，沥干。
2. 把鸡肉用竹签直着串住，每支竹签上串2或3条肉。
3. 花生酱先用温水调稀一点，再加入其他的调味料（2）调匀做成沙茶烤肉酱料。
4. 烤箱预热至200～220℃，把鸡肉串放在烤架上，再放入烤箱中烤6～8分钟。
5. 取出鸡肉串，涂上沙茶酱料，翻面后再放入烤箱中烤3～4分钟，见鸡肉已微焦黄即可。

Ingredients:
2 boneless chicken legs,
12 bamboo skewers

Seasonings:
(1) 1 tbsp soy sauce, 1 tbsp wine,
    1 tsp sugar, 2 tbsp water
(2) 1 tbsp peanut butter, 2 tbsp warm water,
    2 tbsp sha-cha sauce, 1/2 tbsp soy sauce

Procedures:
1. Trim chicken legs, remove all skin and soft bones. You may use chicken breast if you like. Cut chicken to long strips, the size about 2 cm wide. Marinate with seasonings (1) for 10 minutes. Drain.
2. Pierce 2～3 pieces of chicken meat into one skewer.
3. Mix seasonings (2) evenly to make the Bar-B-Q sauce.
4. Preheat oven to 200~220℃, arrange chicken skewers on a rack, bake chicken for 6～8 minutes.
5. Remove baking rack, brush Bar-B-Q sauce over chicken, turn chicken over, continue to bake for 3～4 minutes more until chicken are golden brown.

去骨鸡腿

梅干菜蒸鸡球
Steamed Chicken with Fermented Cabbage

材料：
去骨鸡腿2只、茭白笋2棵、
新鲜豆皮2片、葱1根（切小段）、
大蒜2粒（剁碎）、嫩梅干菜1杯

调味料：
（1）酱油1大匙、盐 1/4 茶匙、水2大匙、淀粉2茶匙
（2）水4大匙、酒 1/2 大匙、糖 1/2 茶匙、麻油 1/2 茶匙

做法：
1. 用刀在鸡腿的表面上剁些刀口，切成约2.5厘米的小块，拌上调味料（1）。
2. 梅干菜快速冲洗一下，略剁碎一点。
3. 茭白笋切成长条块；豆皮切成宽条，铺在蒸盘上。鸡肉和茭白笋拌和，放在豆皮上。
4. 起油锅，用2大匙油爆香蒜末和葱段，放下梅干菜再炒至香气透出。
5. 加入调味料（2）煮滚，再淋在鸡腿和茭白上，上锅蒸20～25分钟，至鸡肉已熟即可取出。

Ingredients:
2 de-boned chicken legs, 2 jiao-bai bamboo shoots,
2 pieces fresh bean curd pack,
1 stalk green onion (sectioned),
2 cloves garlic (chopped),
1 cup fermented cabbage

Seasonings:
(1) 1 tbsp soy sauce, 1/4 tsp salt,
    2 tbsp water, 2 tsp cornstarch
(2) 4 tbsp water, 1/2 tbsp wine,
    1/2 tsp sugar, 1/2 tsp sesame oil

Procedures:
1. Make a few cuts on the meat side, then cut into 2.5 cm cubes. Mix with seasonings (1).
2. Rinse fermented cabbage quickly, drain and cut it shorter.
3. Trim jiao-bai bamboo shoot, then cut into pieces; Cut bean curd pack into wide stripes, arrange on a plate. Combine chicken and bamboo shoot, put on top of bean curd pack.
4. Heat 2 tablespoons of oil to sauté garlic and green onion, when fragrant, add fermented cabbage, sauté until fragrant again.
5. Pour seasonings (2) in, bring to a boil. Pour over chicken. Steam over high heat for about 20 ~ 25 minutes until chicken is cooked.

去骨鸡腿

京都子鸡
Chicken with Jing-Du Sauce

材料：
去骨鸡腿2只、洋葱丝1杯、白芝麻1大匙

调味料：
（1）酱油2大匙、酒1大匙、淀粉2大匙、面粉2大匙、水3大匙
（2）番茄酱2大匙、辣酱油1大匙、A1牛排酱1 $1/2$ 大匙、清水4大匙、糖 $1/2$ 大匙、麻油 $1/4$ 茶匙

做法：
1. 将鸡腿剁成长条块，用调味料（1）拌匀，腌30分钟以上。
2. 用少量油炒熟洋葱丝，加盐调味，盛放在盘内。
3. 烧热4杯炸油，放入鸡腿，以中火炸2分钟，捞出。
4. 再将油烧热，放入鸡肉，以大火炸脆外表即可捞起，沥干油渍。
5. 油倒出，再把调匀的调味料（2）倒入锅中，炒煮至滚。
6. 放入鸡腿和酱汁炒匀，盛放至洋葱上，再撒下白芝麻。

Ingredients:
2 boneless chicken legs,
1 cup shredded onion,
1 tbsp sesame seeds

Seasonings:
(1) 2 tbsp soy sauce, 1 tbsp wine,
   2 tbsp cornstarch, 2 tbsp flour, 3 tbsp water
(2) 2 tbsp ketchup, 1 tbsp Worcestershire sauce,
   1 $1/2$ tbsp A1 steak sauce, 4 tbsp water, $1/2$ tbsp sugar, $1/4$ tsp sesame oil

Procedures:
1. Cut chicken to wide strips, marinate with seasonings (1) for 30 minutes.
2. Sauté onion shreds, season with a little of salt, remove to a serving plate.
3. Heat 4 cups of oil to deep fry chicken over medium heat for 2 minutes. Drain.
4. Reheat oil, deep-fry chicken over high heat until chicken turn golden brown and crispy.
5. Pour oil away, add seasonings (2) in, bring to a boil. Return chicken to sauce, mix evenly. Put on top of onion, sprinkle sesame seeds over chicken.

去骨鸡腿

香酥鸡排堡
Chicken Hamburgers

材料:
鸡腿上半部4只、生菜叶4片、番茄片4片、汉堡面包4个、洋葱末2大匙、酸黄瓜末1大匙、美奶滋3大匙、炸鸡粉1杯、起酥粉1杯

调味料:
盐1茶匙、胡椒粉少许、鸡蛋1个

做法:
1. 选用筋较少的鸡腿上半部，用刀在表面上斩剁数刀，再用调味料拌匀，腌20分钟。
2. 将1/4杯的炸鸡粉和1/4杯的酥炸粉加2/3杯水调成稀糊状。再把剩下的两种粉料各3/4杯混合均匀。
3. 洋葱末、酸黄瓜末和美奶滋调匀，涂在烤热的面包上，再分别放上切条的生菜叶和番茄片。
4. 鸡腿先蘸满一层粉料，再在面糊中蘸一下，再蘸上粉料。
5. 炸油5杯烧至8分热，放下鸡腿，改成小火炸熟，约炸2分半钟，取出。
6. 油再烧热，放下鸡排，用大火炸30秒钟，捞出，沥净油渍，放在汉堡面包上，夹成鸡排堡。

Ingredients:
4 chicken thighs, 4 pieces lettuce leaf,
4 slices tomato, 4 hamburger buns, 2 tbsp chopped onion,
1 tbsp chopped pickles, 3 tbsp mayonnaise,
1 cup fried chicken powder, 1 cup cheese crispy powder

Seasonings:
1 tsp salt, a pinch of pepper, 1 egg

Procedures:
1. Chop few times on the meat side of chicken thigh, then marinate with seasonings for 20 minutes.
2. Mix 1/4 cup of fried chicken powder and 1/4 cup of cheese crispy powder with 2/3 cup of water to make a paste. Then mix remaining powder together.
3. Mix chopped onion and pickles with mayonnaise, then spread over baked hamburger buns. Place lettuce and tomato slice on each halve.
4. Coat chicken thighs with powder, then dip in paste, and then coat with powder again.
5. Heat 5 cups of oil to 160℃, deep-fry chicken over low heat until done, about 2.5 minutes. Remove chicken.
6. Reheat oil, deep-fry over high for 30 seconds. Remove and drain off oil, place on buns to make chicken hamburgers.

去骨鸡腿

Steamed Chicken Rice Pudding

去骨鸡腿
121

八宝鸡排饭
Steamed Chicken Rice Pudding

材料：
鸡腿2只、虾米2大匙、香菇2朵、
笋丁2大匙、红葱头3～4片、长糯米2杯

调味料：
酱油3大匙、酒1大匙、糖1茶匙、胡椒粉 1/4 茶匙

做法：
1. 鸡腿剔除大骨，用刀在白筋上剁一些刀口，用调味料拌腌15～20分钟。
2. 烧热4大匙油，鸡腿的皮面朝下放入锅中，煎黄鸡皮，翻面再以小火煎至鸡腿较定型（约2分钟）。取出切成宽条，排在扣碗底部。
3. 长糯米洗净、加水1 1/2 杯，煮成糯米饭。
4. 虾米泡软、摘好，略切一下；香菇泡软，切成小丁；红葱头切片。
5. 起油锅，用2大匙油炒香红葱片，盛出。再放入香菇和虾米爆香，加入笋丁，淋下剩余的腌鸡料和水3大匙，煮滚后关火。
6. 放入糯米饭和红葱酥拌匀，再把糯米饭填塞到碗中，用铝箔纸或保鲜膜封住碗口。
7. 把鸡排饭放入蒸锅中蒸30～40分钟，取出，倒扣在盘中。

Ingredients:
2 chicken legs, 2 tbsp dried shrimp, 2 shitake mushrooms,
2 tbsp diced bamboo shoot, 3 ~ 4 pieces shallot, 2 cups glutinous rice

Seasonings:
3 tbsp soy sauce, 1 tbsp wine, 1 tsp sugar, 1/4 tsp pepper

Procedures:
1. Remove bones from chicken, then make a few cuts on meat side to cut off the white tendons. Marinate with seasonings for 15 ~ 20 minutes.
2. Heat 4 tablespoons of oil to fry chicken (with skin side down). Turn it over when browned. Continue to fry for 2 minutes until chicken meat is firmed. Remove and cut to wide strips, arrange on a bowl.
3. Rinse the rice, add 1 1/2 cups of water to cook the rice in a rice cooker.
4. Soak dried shrimps to soft, then cut to smaller size. Soak and dice mushrooms. Slice shallots.
5. Sauté shallot with 2 tablespoons of oil until fragrant, remove shallot. Add mushroom and dried shrimp, stir until fragrant. Add bamboo shoot, the remaining seasonings and 3 tablespoons of water. Turn off the heat when boils.
6. Add cooked rice and shallot in, mix evenly. Put rice on top of the chicken, seal the bowl with a piece of foil.
7. Steam chicken rice pudding for 30 ~ 40 minutes. Remove and reverse on a serving plate.

去骨鸡腿

Chicken Salad with Spicy Sauce

去骨鸡腿
123

椒麻鸡
Chicken Salad with Spicy Sauce

材料：
鸡腿1只（约150克）、圆白菜丝1杯、
葱2根、姜2片、红辣椒1个、大蒜2粒、香菜1～2根

调味料：
（1）淡色酱油 1/2 大匙、酒1大匙、胡椒粉少许
（2）酱油2大匙、醋1大匙、柠檬汁1大匙、糖2大匙、盐少许、花椒粉 1/3 茶匙、麻油 1/2 大匙

做法：
1. 鸡腿去骨，用刀子将白筋部分斩剁一下，并将肉厚的地方片开。把1根葱和姜拍碎，加入调味料（1）和鸡腿拌匀，腌20分钟。
2. 圆白菜丝洗净，用冰水泡10分钟，沥干，并以纸巾吸干水分，放在餐盘中。
3. 另一根葱切成细葱末；大蒜磨成蒜泥；红辣椒去籽，切碎；香菜也切碎。碗中将调味料（2）调匀，放入葱末和蒜泥。
4. 锅中将3杯油烧至7分热，放入鸡腿先大火炸10秒钟，转成小火再慢慢炸至熟，捞出。
5. 将油再烧热，放下鸡肉，以大火炸15秒钟，待酥脆时捞出，切成条，排在圆白菜上。
6. 淋下调味汁，撒下红椒末和香菜段。上桌后拌匀。

Ingredients:
1 chicken leg (about 150g.), 1 cup cabbage shreds,
2 stalks green onion, 2 slices ginger, 1 red chili, 2 cloves garlic, 1～2 stalks cilantro

Seasonings:
(1) 1/2 tbsp light colored soy sauce, 1 tbsp wine, a pinch of pepper
(2) 2 tbsp soy sauce, 1 tbsp vinegar, 1 tbsp lemon juice, 2 tbsp sugar, a little of salt, 1/3 tsp brown pepper power, 1/2 tbsp sesame oil.

Procedures:
1. Remove bones from chicken leg, cut through the tendons, split the thick part of meat to make the whole leg evenly. Crush 1 stalk of green onion and ginger, mix with seasonings (1) to marinate chicken for 20 minutes.
2. Rinse cabbage shreds, soak in ice water for 10 minutes. Drain and pat it dry with paper towel, place on a serving plate.
3. Chop green onion finely; smash garlic; remove seeds from red chili then chop it; chop cilantro; mix seasonings (2), green onion and garlic in a bowl to make the sauce.
4. Heat 3 cups of oil to 140℃, deep fry chicken over high heat for 10 seconds, turn to low heat, continue to deep fry until done. Drain.
5. Reheat oil, deep fry chicken again over high heat for only 15 seconds. Drain. Cut into strips, arrange on top of cabbage.
6. Pour sauce over chicken, sprinkle red chili and cilantro on top, serve. Mix evenly before eating.

去骨鸡腿

红烩鸡腿排
Chicken Legs with Tomato Sauce

材料：
鸡腿2只、洋菇10朵、洋葱 1/2 个、
红葱末 1/2 大匙、罐头去皮番茄5个、面粉 2～3大匙

调味料：
（1）酒1大匙、酱油1大匙、盐 1/4 茶匙、胡椒粉少许、水2大匙
（2）酒1大匙、番茄酱1大匙、辣酱油 1/2 大匙、糖 1/2 茶匙、水2杯

做法：
1. 鸡腿剔除腿骨，但要留下关节骨。在鸡肉面上剁上一些刀口，用调味料（1）腌10～15分钟。
2. 洋菇切片；洋葱切丝；去皮番茄切大块，罐头汤汁留用。
3. 鸡腿蘸上面粉，用3大匙热油煎黄两面，夹出鸡腿。
4. 用锅中的余油炒香洋葱丝和红葱末，再加入洋菇同炒，淋下酒、番茄汁和其他的调味料（2）。
5. 汤汁煮滚后再放入鸡腿和番茄块，以中小火烧煮10～12分钟至熟。
6. 夹出鸡腿，切成宽条，排入盘中，淋下番茄洋菇酱汁。

※如果汤汁不够浓稠时，可以将少许面粉筛入汤汁中，使汤汁浓稠，再淋到鸡腿排上。

Ingredients:
2 chicken legs, 10 mushrooms, $1/2$ onion,
$1/2$ tbsp chopped shallot, 5 tomatoes (canned), 2～3 tbsp flour

Seasonings:
(1) 1 tbsp soy sauce, 1 tbsp wine, $1/4$ tsp salt, a pinch of pepper, 2 tbsp water
(2) 1 tbsp wine, 1 tbsp ketchup, $1/2$ tbsp Worcestershire sauce, $1/2$ tsp sugar, 2 cups water

Procedures:
1. Remove bones from chicken legs, but keep the joint bone in the end of drum stick; chop a few cuts on the meat side; marinate with seasonings (1) for 10～15 minutes.
2. Slice mushrooms; shred onion; cut canned tomatoes to large pieces, keep the tomato juice.
3. Coat chicken legs with flour; fry both sides to golden brown; remove.
4. Stir-fry onion shreds and shallot until fragrant; add mushroom; stir fry for a while. Add wine, tomato juice, and other seasonings (2), bring to a boil.
5. Add chicken legs and tomatoes, cook over medium-low heat for about 10～12 minutes until chicken legs are cooked.
6. Remove chicken legs, cut into pieces; arrange on a plate, pour tomato sauce over chicken.

※If the sauce is not thick enough, shift in a little of flour to thicken it.

去骨鸡腿

Chicken with Mushrooms in Casserole

※For this dish, you should use chicken of the best quality, so you can get better health benefits.

北菇滑鸡煲
Chicken with Mushrooms in Casserole

材料：

去骨鸡腿2只、香菇6～8朵、葱2根、姜5～6小片、西生菜 1/2 球、淀粉1大匙

蒸香菇料：

酱油1大匙、葱1根、油1大匙、糖1茶匙、泡香菇水1杯

调味料：

（1）酱油 1 1/2 大匙、水2大匙、淀粉 1/2 大匙

（2）酒1大匙、酱油 1/2 大匙、蚝油1大匙、糖 1/2 茶匙、水 2/3 杯、麻油 1/2 茶匙、淀粉1茶匙

做法：

1. 鸡腿切成2厘米大小，用调味料（1）拌匀，腌20～30分钟。
2. 香菇泡软，加少许淀粉抓洗一下，再用水冲洗干净，放在大碗中，加蒸香菇料蒸15分钟。待稍凉后取出，切成2或3片。
3. 西生菜切成大片，在滚水中一烫即捞出，盛放在沙锅中。
4. 水再煮滚，放下鸡块烫煮至五六分熟，捞出。
5. 起油锅用2大匙油爆香葱段和姜片，放入香菇再炒香，加入鸡块和调匀的调味料（2）同煮。
6. 煮至汤汁浓稠且鸡肉已熟，全部倒在生菜上，再煮滚即可上桌。

※北菇指的是在广东北江一带产的香菇，是质量较好、厚而有香气的冬菇。

Ingredients:

2 boneless chicken leg, 6 ~ 8 pieces shitake mushroom,
2 stalks green onion, 5 ~ 6 ginger slices, 1/2 head lettuce, 1 tbsp cornstarch

To steam mushrooms:

1 tbsp soy sauce, 1 stalk green onion, 1 tbsp oil,
1 tsp sugar, 1 cup water (from soaked mushrooms)

Seasonings:

(1) 1 1/2 tbsp soy sauce, 2 tbsp water, 1/2 tbsp cornstarch

(2) 1 tbsp wine, 1/2 tbsp soy sauce, 1 tbsp oyster sauce, 1/2 tsp sugar, 2/3 cup water, 1/2 tsp sesame oil, 1 tsp cornstarch

Procedures:

1. Cut chicken to 2cm pieces, marinate with seasonings (1) for 20 ~ 30 minutes.
2. Soak mushrooms to soft, remove from water. Mix mushrooms with cornstarch, stir for a while to clean the mushrooms, rinse with water, put in a bowl. Add steaming ingredients, steam mushrooms for 15 minutes. Remove and then cut into 2 or 3 pieces when it cools.
3. Cut lettuce to large pieces, blanch quickly, remove and then place in a casserole.
4. Bring water to a boil again, blanch chicken to half cooked, remove chicken.
5. Heat 2 tablespoons of oil to sauté green onion and ginger. Add mushrooms, continue to sauté until fragrant. Add chicken and mixed seasonings (2) in.
6. Cook until chicken is done, pour to the casserole. Serve after boils again.

去骨鸡腿

Crispy Chicken with Spices

去骨鸡腿
129

去骨盐酥鸡
Crispy Chicken with Spices

材料：
去骨鸡腿2只、葱2根、姜2片、
番薯粉 2/3 杯、面粉1大匙、九层塔3～4枝

调味料：
（1）酒1大匙、盐 1/4 茶匙、胡椒粉少许、蛋黄1个
（2）五香粉 1/2 茶匙、白胡椒粉 1/2 茶匙、盐1茶匙、大蒜粉1茶匙

做法：
1. 将鸡腿剁成3厘米大小的鸡块，放在大碗中。加入拍碎的葱、姜及调味料（1）一起拌匀，腌30分钟以上。
2. 番薯粉和面粉拌匀后用来蘸裹鸡块，尽量使每块鸡肉均裹上番薯粉，放置3～5分钟。
3. 锅中把4杯炸油烧至8分热，放入鸡肉，以中小火炸至9分熟（约1.5分钟），先捞出鸡块。
4. 将油再烧热，重新放下鸡块，以大火再炸15～20秒钟，至外表酥脆，捞出、沥净油渍。
5. 放下摘好的九层塔炸一下，变略透明时即捞出，和鸡块略拌，撒下调匀的调味料（2）即可。

Ingredients:
2 boneless chicken legs, 2 stalks green onion, 2 slices ginger,
2/3 cup sweet potato powder, 1 tbsp flour, 3～4 stalks basil

Seasonings:
(1) 1 tbsp wine, 1/4 tsp salt, a pinch of pepper, 1 egg yolk
(2) 1/2 tsp five spice powder, 1/2 tsp white pepper, 1 tsp salt, 1 tsp garlic powder

Procedures:
1. Cut chicken leg into 3cm pieces. Mix evenly with crushed green onions and ginger, add seasonings (1) in, marinate for at least 30 minutes.
2. Mix sweet potato powder with flour. Coat chicken with the powder, set aside for 3～5 minutes.
3. Heat 4 cups of oil to 160℃, deep-fry chicken over medium heat for about 1.5 minutes until almost done. Remove chicken.
4. Reheat the oil and deep-fry chicken again over high heat until the chicken becomes very crispy, about 15～20 seconds. Drain.
5. Deep-fry basil leaves, remove quickly when the color has changed, mix with chicken. Sprinkle mixed seasonings (2) over chicken. Serve hot.

去骨鸡腿

Steamed Chicken Packed with Lotus Leaves

去骨鸡腿
131

荷叶粉蒸白果鸡
Steamed Chicken Packed with Lotus Leaves

材料：
去骨鸡腿2只、香菇4朵、银杏32粒、
干荷叶2张、蒸肉粉1杯（2小盒，粗细各一盒）

调味料：
葱2根、姜2片、酱油2大匙、糖1大匙、酒1大匙、
油3大匙、甜面酱2茶匙、辣豆瓣酱1茶匙、水4大匙

做法：
1. 鸡腿每只分切成4大块，洗净，擦干水分。
2. 葱、姜拍碎，加入调味料拌匀，放入鸡块，再抓拌均匀，腌约1小时。
3. 香菇泡软，切成两半；白果剥壳、再剥去褐色薄膜，如果用真空包装的银杏，要用水冲洗数次。
4. 干荷叶用温水泡软、轻轻刷净、擦干，剪掉中间的硬梗部分，每张再分切成4小张。
5. 把葱、姜挑出，加入蒸肉粉拌匀，再包入荷叶中（要尽量多蘸蒸肉粉），同时每包中再放入1片香菇、4粒银杏，包成长方包，排列在蒸碗中。
6. 以中小火蒸1.5小时以上，至鸡腿已软烂。

※蒸的时间依不同鸡种而决定。

Ingredients:
2 boneless chicken legs, 4 shitake mushrooms, 32 ginkgo nuts,
2 pieces dried lotus leaf, 1 cup flavored rice powder

Marinades:
2 stalks green onion, 2 slices ginger, 2 tbsp soy sauce, 1 tbsp sugar,
1 tbsp wine, 3 tbsp oil, 2 tsp soy bean paste, 1 tsp hot bean paste, 4 tbsp water

Procedures:
1. Cut each chicken leg to 4 pieces, Rinse and pat dry.
2. Crush green onion and ginger, mix well with seasonings, then put chicken in, marinate for 1 hour.
3. Soak shitake mushrooms to soft, then halve each one. Peel ginkgo nuts and remove the red membrane. Rinse for several times if you use those ginkgo nuts which are packed in airtight bags.
4. Soak dried lotus leaves in warm water until soft, brush gently. Wipe dry, cut off the stem part, then divide each one to 4 pieces.
5. Remove green onion and ginger, then mix rice powder with chicken, cover chicken with rice powder as much as possible. Wrap one piece of chicken, one piece of shitake mushroom and 4 ginkgo nuts with one piece of lotus leaf to form a rectangle shape. Place in a bowl.
6. Steam chicken over medium-low heat for about 1.5 hours until chicken are tender enough. Serve hot.

※The steaming time is depend on what kind of chicken you are useing.

亲子丼（两碗份）
Rice Cover with Chicken & Egg Sauce <2 servings>

材料：
去骨鸡腿肉1只、新鲜香菇3朵、
洋葱丝 1/2 杯、葱1根、鸡蛋2个、白饭2碗、葱花少许

调味料：
（1）酒2茶匙、盐 1/3 茶匙、胡椒粉少许
（2）香菇酱油2大匙、味霖1大匙、盐少许、清汤1 1/2 杯

做法：
1. 鸡腿切成6～7小块，均匀地撒上调味料（1），腌10分钟。
2. 新鲜香菇切条；葱切段；鸡蛋打散。
3. 起油锅，用2大匙油将洋葱丝和葱段炒香，加入调味料（2）煮滚。
4. 放入鸡块和香菇，以中火续煮2～3分钟至鸡腿肉已熟。
5. 在汤汁滚动处淋下蛋汁，成为片状的蛋花，见蛋汁几乎凝固时关火，撒下葱花，全部淋在热的白饭上。

※淋蛋汁时不要淋在一个地方，手要绕着锅子转一圈，使蛋汁均匀淋下成蛋片状。

Ingredients:
1 boneless chicken leg,
3 fresh shitake mushrooms,
1/2 cup onion shreds,
1 stalk green onion, 2 eggs,
2 cups cooked rice, a little of
chopped green onion

Seasonings:
(1) 2 tsp wine, 1/3 tsp salt, pinch of pepper
(2) 2 tbsp mushroom soy sauce, 1 tbsp mirin,
salt to taste, 1 1/2 cups soup stock

Procedures:
1. Cut chicken into 6～7 pieces, mix with seasonings (1) for 10 minutes.
2. Cut mushrooms to stripes; section green onion; beat the eggs.
3. Heat 2 tablespoons of oil to sauté onion and green onion, add seasonings (2), bring to a boil.
4. Add chicken and mushroom in, cook over medium heat for 2~3 minutes until chicken is cooked.
5. Pour egg mixture to soup while soup is boiling. Turn off the heat when egg is almost solidified. Add chopped green onion in, pour it over cooked rice.

※The egg should be poured around the pot instead of at one spot, so that it will be distributed evenly.

去骨鸡腿

鸡肉丸子汤
Chicken Meatballs Soup

材料：
鸡腿2只、熟笋小丁2大匙、葱1根、姜2片、菠菜2~3根、番茄1个、水5杯

调味料：
（1）盐 1/2 茶匙、鸡蛋1个、淀粉2茶匙、麻油 1/4 茶匙、白胡椒粉少许
（2）盐、胡椒粉、麻油各适量调味

做法：
1. 鸡腿去骨、去皮后剁碎，放入大碗中；番茄切块；菠菜切段。
2. 葱、姜拍一下，加入3大匙水和 1/2 大匙酒抓一下浸泡5分钟，做成葱姜酒汁。
3. 鸡腿中先加盐和葱姜酒汁，搅拌至有黏性，再加入笋丁和其余的调味料（1），搅拌均匀。
4. 锅中用1大匙油炒一下番茄，加水煮滚。
5. 将鸡肉挤成球形，放入汤中，以中小火煮至鸡球浮起，改小火再煮2分钟。
6. 加盐调味，最后放下菠菜，再煮滚即可关火，加入麻油和胡椒粉。

※鸡肉不要剁的太细，有小颗的鸡肉粒在丸子里才更有口感。

Ingredients:
2 chicken legs, 2 tbsp diced bamboo shoot (cooked),
1 stalk green onion, 2 slices ginger,
2 ~ 3 stalks spinach, 1 tomato, 5 cups water

Seasonings:
(1) 1/2 tsp salt, 1 egg, 2 tsp cornstarch, 1/4 tsp sesame oil, a pinch of pepper
(2) salt, pepper and sesame oil to taste

Procedures:
1. Remove bones and skin from chicken, cut to small pieces and then chop for a while. Put in a large bowl. Cut tomato to pieces. Cut spinach to sections.
2. Crush green onion and ginger, soak in 3 tablespoons of water and 1/2 tablespoon of wine for 5 minutes.
3. Mix chicken with salt and green onion wine juice, stir until sticky. Add bamboo shoot and rest of the seasonings (1), stir evenly.
4. Stir-fry tomato with 1 tablespoon of oil, add water in, bring to a boil.
5. Make balls with chicken mixture, put into soup, cook over medium low heat until balls float up. Cook for 2 more minutes over low heat.
6. Season with salt, add spinach, turn off the heat when soup boils again. Add pepper and sesame oil at last.

※Having some small pieces of chicken helps to give the meatballs good texture. Therefore, do not over-mince the chicken meat.

去骨鸡腿

干锅鸡
Stir-fried Chicken in Casserole

材料：
去骨鸡腿1只、鸡肫2个、鸡肝1个、
青椒 1/2 个、红辣椒1个、新鲜香菇3朵、
姜片10小片、大蒜4粒、芹菜2枝
调味料：
（1）酒1大匙、胡椒粉少许
（2）酱油2大匙、酒1大匙、糖 1/3 茶匙、水 1/4 杯、花椒粉 1/4 茶匙
做法：
1. 鸡腿切小块；鸡肫、鸡肝分别切片；三种材料用调味料（1）拌腌一下。
2. 青椒切条；红辣椒切圈；新鲜香菇切片；大蒜切片；芹菜切段。
3. 起油锅把2杯炸油烧热，放下鸡肉、鸡肫和鸡肝，以大火炸至熟，捞出，放入一个加热的沙锅或酒精锅中。
4. 放下大蒜、青椒、红椒和鲜香菇，再继续拌炒，边炒边加入调味料（2），要不断翻炒以免黏住锅底。
5. 最后加入芹菜段，再翻炒一下即可关火上桌。

※用酒精锅的话可以上桌后再开火来炒热。

Ingredients:
1 boneless chicken leg, 2 gizzard,
1 chicken liver, 1/2 green pepper,
1 red chili, 3 pieces fresh mushroom,
10 slices ginger, 4 clovers garlic, 2 stalks celery
Seasonings:
(1) 1 tbsp wine, pinch of pepper
(2) 2 tbsp soy sauce, 1 tbsp wine,
   1/3 tsp sugar, 1/4 cup water, 1/4 tsp brown pepper powder
Procedures:
1. Cut chicken to small pieces; slice gizzard and liver; mix three ingredients with seasonings (1).
2. Cut green onion to stripes; cut red chili to circles; slice mushrooms; slice garlic; cut celery to sections.
3. Heat 2 cups of oil to deep-fry chicken, gizzard and liver over high heat until done. Remove to a heated casserole or a pot.
4. Add in garlic, green & red chilies, and mushrooms, continue to stir-fry. Add seasonings (2) while stirring. Keep on stirring until all mixed.
5. Add celery in at last, stir-fry briskly. Serve hot.

※You may use a gas stove and stir-fry this dish on table while eating.

去骨鸡腿

咸鱼鸡粒豆腐煲
Chicken with Salty Fish & Tofu in Casserole

材料：
去骨鸡腿1只、咸鱼80克、
豆腐2方块、葱1根、姜片6~8小片、葱花1大匙

调味料：
（1）酱油 1/2 大匙、麻油1茶匙、胡椒粉少许、淀粉1茶匙、水1大匙
（2）酒1大匙、蚝油1大匙、高汤1杯、糖1茶匙、淀粉水适量

做法：
1. 在鸡腿的表面上剁一些刀口，再切成2厘米大小，用腌鸡料拌匀腌20分钟。
2. 咸鱼切成如黄豆般小粒，用油炒炸至酥，盛出。
3. 豆腐切成3厘米大小，用4杯热油炸黄外表，捞出，放入沙锅中。
4. 油倒出，仅留 1/2 杯左右，将鸡粒快速过油，炒至8分熟，捞出。
5. 仅用约1大匙油爆香葱段和姜片，淋下酒、蚝油、糖和高汤，煮滚后倒入沙锅中。
6. 放下鸡粒和咸鱼丁，轻轻拌和，一起再煮约1分钟，勾芡后再加入葱花即可关火、上桌。

Ingredients:

1 boneless chicken leg,
80g. salty fish, 2 pieces tofu,
1 stalk green onion, 6~8 pieces ginger,
1 tbsp chopped green onion

Seasonings:

(1) 1/2 tbsp soy sauce, 1 tsp sesame oil,
    pinch of pepper, 1 tsp cornstarch, 1 tbsp water
(2) 1 tbsp wine, 1 tbsp oyster sauce,
    1 cup soup stock, 1 tsp sugar, cornstarch paste

Seasonings:

1. Make a few cuts on the meat side of chicken, then cut into 2cm pieces. Marinate with seasonings (1) for 20 minutes.
2. Cut salty fish to small soybean size. Fry with hot oil until crispy. Remove.
3. Cut tofu to 3cm squares. Deep-fry with hot oil until golden brown, remove and place in a casserole.
4. Pour oil away, keep only 1/2 cup in wok, fry chicken quickly until almost done, remove.
5. Keep only 1 tablespoon of oil to sauté green onion sections and ginger, add wine, oyster sauce, sugar and soup stock in, bring to a boil, pour into the casserole.
6. Put chicken and salty fish in, cook for about 1 minute. Thicken the soup with cornstarch paste, add chopped green onion in. Turn off the heat, serve hot.

Stir-fried Noodles with Chicken in Sha-Cha Sauce

去骨鸡腿

沙茶鸡肉炒面（四人份）
Stir-fried Noodles with Chicken in Sha-Cha Sauce (4 servings)

材料：
鸡腿2只、香菇5朵、青花椰菜1棵、葱2根、姜片3～4片、油面600克

调味料：
（1）酱油1 1/2 大匙、水2大匙、淀粉1茶匙
（2）沙茶酱3大匙、酱油1大匙、酒 1/2 大匙、盐 1/2 茶匙、糖 1/2 茶匙、清汤1杯

做法：
1. 鸡腿去骨后在表面上轻轻斩剁数刀，再切成块，用调味料（1）拌匀，腌20分钟。
2. 香菇泡软、切成片；青花椰菜摘成小朵，用热水氽烫一下，捞出、冲凉。
3. 锅中烧热4大匙油，把鸡肉炒至8分熟，盛出。
4. 再将葱段、姜片和香菇片放入锅中炒香，先放下沙茶酱，再加入其他的调味料（2）炒匀。
5. 放下鸡球、青花椰菜和油面，挑拌均匀。盖上锅盖焖煮1分钟即可，拌匀盛出。

Ingredients:
2 chicken legs, 5 shitake mushrooms, 1 broccoli, 2 stalks green onion, 600g. cooked noodles

Seasonings:
(1) 1 1/2 tbsp soy sauce, 2 tbsp water, 1 tsp cornstarch
(2) 3 tbsp sha-cha sauce, 1 tbsp soy sauce, 1/2 tbsp wine, 1/2 tsp salt, 1/2 tsp sugar, 1 cup soup stock

Procedures:
1. Remove bones from chicken leg, chop a few cuts on the meat side, then cut to small pieces. Marinate with seasonings (1) for 20 minutes.
2. Soak shitake mushrooms to soft, then slice it; Trim broccoli to small pieces, blanch and then rinse, drain.
3. Heat 4 tablespoons of oil to stir-fry chicken until almost cooked. Drain.
4. Add green onion, ginger and mushrooms in, sauté until fragrant. Add sha-cha sauce and the rest of seasonings (2), stir evenly.
5. Add chicken, broccoli and noodles, stir-fry until evenly mixed. Cover the lid, cook for about 1 minute over low heat. Remove and serve.

Part II 带骨鸡腿 Bone-in Chicken Legs

带着鸡腿骨一起烹调的菜所使用的烹调方法是很广泛的，它的特色是以腿骨来保持鸡腿的形状；同时连骨一起炖煮时能使滋味更浓厚，有些人还特别喜欢啃骨头。

目前，在市场上可以买到分割好的鸡腿，多半是肉鸡和仿土鸡（土鸡较少分割出售），一般来说，仿土鸡比肉鸡大，关节处呈现黑灰色，肉鸡则为白色（图1）。肉鸡较小而肉质软嫩，适合炸、烤和烧卤。仿土鸡肉质有弹性，而同为仿土鸡，公鸡的腿比母鸡的腿大（图2）、肉质较粗，比较适合煮汤、红烧和用炖、焖的方式来烹调。做菜之前，我们对材料要先进行了解，按照烹调方式来买适合的材料是最重要的。

图1

Bone-in chicken legs are eligible for many cooking methods. The bone not only maintains the shape of the meat, but also adds additional flavor during the cooking process.

Separated chicken legs that are available in markets are mostly broiler and Simulate native chicken, since native chicken are mostly sold whole. Generally the legs of simulate native chicken are larger than broiler's, and the color of their ankle joints are dark gray (pic 1). Broiler are also more tender, which makes them suitable for deep-frying, baking and stewed. Simulate native cock are larger than the hens (pic 2) and their meat is also firmer, suitable for making soup, stewed or braised. It is very important to understand the ingredients so that you can choose the most appropriate cooking method.

图2

带骨鸡腿的烹调要领：

带骨鸡腿烹调时很简单，一种是剁成块（图3）；另一种则是大块的——保持整个原形或剁成棒棒腿（图4）和上大腿两块。

整只鸡腿的腌泡或炸都需要较长的时间，因此腌鸡腿之前，用叉子在肉厚的地方叉几下（图5）；或在内侧肉面沿着骨头划一道刀口，都是容易入味且容易熟的方法。

当然，在后面介绍的一些菜式，并不是一定要用鸡腿来做，也可以用整只鸡剁成块来烧煮，只是现在的人都讲究口感，较长时间烧出来的鸡腿，肉质是胜过鸡胸的。至于煮鸡的时间长短，一般是依照鸡的品种、剁的大小块而有差异，肉鸡只需15～20分钟，仿土鸡要保持嫩度的话，在20～25分钟，熟了即可，或是烧50～60分钟，使它更入味；这当然就是依照个人喜欢的口感来决定啦！

Cooking tips:

Bone-in chicken legs are either chopped into pieces (pic 3) or cooked in large chunks (keep in whole piece or separate into drum stick and thigh). It takes longer to marinate or deep-fry whole chicken legs. Poking the thicker part of the meat with a fork (pic 5) or cutting a small score along the bone can help speed up the process.

You may substitute chicken legs with whole chicken for some dishes in this book. The cooking time for chicken not only depends on the size but also the breed of chicken you are cooking. Broiler takes only 15～20 minutes; however, simulate native chicken takes 20～25 minutes to keep the tenderness, or up to 50～60 minutes to make it more tasteful. It is all up to your personal preference.

图3

图4

图5

轻烟熏鸡腿
Smoked Chicken

材料：
鸡腿3只、鸡翅4只、铝箔纸1张、少许麻油

调味料：
酱油5大匙、酒2大匙、冰糖 1/2 大匙、葱1根、姜2片、大料1粒、桂皮1段、盐1茶匙、水6杯

熏料：
红茶叶2大匙、面粉3大匙、红糖3大匙

做法：
1. 鸡腿和鸡翅洗净，擦干。
2. 调味料放汤锅中先煮至滚，再以小火煮10分钟。
3. 放入鸡腿，先煮5分钟后加入鸡翅，再煮10分钟，关火浸泡30分钟，取出放凉。
4. 一个耐烧的炒锅中先铺一张铝箔纸，放上熏料，架上一个架子（架子上先涂少许油），把鸡腿和鸡翅放在架子上，鸡皮面朝上，盖上锅盖。
5. 开火，待冒烟时计时，熏5分钟，翻面再熏5分钟，关火，再焖5分钟即可取出。
6. 抹上少许麻油，放凉后即可剁成块。

Ingredients:
3 chicken legs, 4 chicken wings, a piece of foil, sesame oil

Seasoning:
5 tbsp soy sauce, 2 tbsp wine, 1/2 tbsp rock sugar, 1 stalk green onion, 2 slices ginger, 1 star anise, 1 piece cinnamon, 1 tsp salt, 6 cups water

Smoking Ingredients:
2 tbsp black tea, 3 tbsp flour, 3 tbsp brown sugar

Procedures:
1. Rinse chicken, then pat the skin to dry.
2. Boil seasonings in a pot over high heat, then simmer for 10 minutes.
3. Put chicken legs in, cook for 5 minutes. Add chicken wings, cook for 10 minutes more. Turn off the heat, soak for 30 minutes. Remove and let cool.
4. In a wok, place a piece of foil on, then arrange those ingredients for smoking in. Put a rock on top (brush some oil over rock), lay chicken parts on top with skin side up. Cover the lid.
5. Turn on the heat, smoke for 5 minutes from the time you saw the smoke. Turn chicken over, smoke for 5 minutes more. Turn off the heat. Let chicken stay for 5 minutes more.
6. Remove chicken, brush sesame oil, set aside until cools. Cut to pieces before serve.

带骨鸡腿

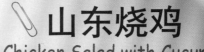

山东烧鸡
Chicken Salad with Cucumber

材料：
仿土鸡腿2只、黄瓜3条、酱油4大匙
蒸鸡料：
花椒粒2大匙、葱2根、姜4片
调味料：
酱油2大匙、醋1 1/2 大匙、大蒜泥1大匙、麻油1大匙、蒸鸡汁2大匙
做法：
1. 鸡腿洗净，擦干水分，用酱油泡1小时，要翻一下面，使颜色均匀。
2. 用2杯热油煎黄鸡的表皮，捞出，沥净油，放在一个有深度的盘中。
3. 将花椒粒和葱段、姜片放在鸡腿上，上锅蒸1小时（用肉鸡时只需蒸30分钟）。
4. 黄瓜拍裂，去除中间的籽，切成段，放入盘中。
5. 将鸡取出，放至凉透后用手撕成粗条，堆放在黄瓜上。
6. 调味料调匀后淋在鸡腿上，临吃前拌匀即可。

Ingredients:
2 simulate native chicken legs,
3 cucumbers, 4 tbsp soy sauce
To steam chicken:
2 tbsp brown peppercorns,
2 stalks green onion, 4 slices ginger
Seasonings:
2 tbsp soy sauce, 1 1/2 tbsp vinegar, 1 tbsp mashed garlic,
1 tbsp sesame oil, 2 tbsp broth from steamed chicken
Procedures:
1. Rinse and pat dry chicken legs. Soak in soy sauce for 1 hour, turn chicken over occasionally to make the color evenly.
2. Heat 2 cups of oil to fry chicken with skin side down. Fry until dark brown. Drain and place on a plate.
3. Place brown peppercorn, green onion, and ginger on top of chicken, steam over high heat for 1 hour (for the meat chicken, 30 minutes is enough).
4. Crush cucumber, remove seeds from the center, cut to sections, place on a plate.
5. Remove chicken. Tear the meat to strips when it cools, place on top of cucumber.
6. Prepare seasoning sauce, pour over chicken. Serve. Mix evenly before eating.

带骨鸡腿

葱油鸡腿
Steamed Chicken Legs with Green Onion

材料：
仿土鸡腿2只、葱丝 1/2 杯、嫩姜丝 1/2 杯

调味料：
葱1根、姜2片、盐1茶匙、酒1大匙

做法：
1. 将鸡腿洗净后，用叉子在鸡腿表面上刺数下，以使腌泡时容易入味。
2. 将盐与酒混合后，再混合拍碎的葱、姜，在鸡腿上抹擦数次后，腌约20分钟。
3. 蒸锅的水烧开，放入鸡腿，用大火蒸约12分钟，熄火后再焖5分钟。
4. 鸡腿取出，放在熟食菜板上剁成块，排列盘中。将切好的葱丝及姜丝（切好后在冷开水中泡3～5分钟）撒在鸡腿上。
5. 将2大匙油烧得极热，淋在葱、姜上，再淋下3大匙蒸鸡的汁，即可上桌。

※鸡块淋过油后，也可将油再倒回炒锅内，与鸡汁混合煮滚，勾薄芡，再浇到鸡腿上。

Ingredients:
2 simulate native chicken legs,
1/2 cup tender ginger shreds,
1/2 cup green onion shreds

Seasonings:
1 stalk green onion, 2 slices ginger,
1 tsp salt, 1 tbsp wine

Procedures:
1. Pierce chicken legs with a fork from the meat side.
2. Crush the green onion and ginger slices, mix with salt and wine. Marinate chicken with it for 20 minutes.
3. Steam chicken for 12 minutes over high heat. Turn off the heat and let the chicken remain in steamer for 5 minutes more.
4. Remove chicken, cut into pieces. Arrange on a platter. Sprinkle green onion and ginger shreds on top (soak them in ice water for 3~5 minutes before hand).
5. Heat 2 tablespoons of oil to very hot, pour oil over the green onion and ginger, also pour 3 tablespoons of chicken broth (remaining from steamed chicken) over, then serve.

※You may pour the oil back to wok after pouring over chicken, reheat it with the chicken broth, thicken with cornstarch paste, then pour over chicken again.

带骨鸡腿

香酥鸡腿
Brown Peppercorn Flavored Crispy Chicken

材料：
棒棒鸡腿6只、葱2根、姜3片、花椒粒2大匙、面粉 1/2 杯

调味料：
盐 3/4 大匙、酒2大匙、酱油2大匙

做法：
1. 在干的锅中，以小火慢慢炒香花椒粒，再加入盐同炒至盐微黄即盛入大碗中。加入拍碎的葱、姜及酒拌匀。
2. 用葱姜料腌鸡腿，放置3小时以上。
3. 放入电锅或蒸锅中蒸1小时以上，至鸡腿够烂时取出。
4. 略放凉后，在鸡腿上抹一层酱油，并拍上一层面粉，随即投入已烧热的炸油中。
5. 以大火、热油把外皮炸成金黄色（可以炸两次至酥黄）。捞出、沥干油渍、装盘。

※ 也可改用整只鸡腿或整只鸡来制作。

Ingredients:
6 chicken drumsticks,
2 stalks green onion, 3 slices ginger,
2 tbsp brown peppercorn, 1/2 cup flour

Seasonings:
3/4 tbsp salt, 2 tbsp wine, 2 tbsp soy sauce

Procedure:
1. Stir-fry the brown peppercorns over low heat in a dried and clean wok until fragrant. Add salt, continue to stir-fry until the salt becomes light brown. Remove to a large bowl. Mix with wine, crushed green onions, and ginger.
2. Marinate chicken legs with brown peppercorn mixture at least for 3 hours.
3. Steam chicken legs in a rice cooker or a steamer over 1 hour until very tender.
4. Remove chicken and let it cool. Brush soy sauce over the legs and then coat with flour.
5. Deep-fry chicken in hot oil until the skin side become golden brown and crispy (it is much better to deep-fry it twice). Drain and arrange on a plate.

※ You may use whole chicken leg or whole chickens instead of drumsticks.

带骨鸡腿

酸辣木耳鸡
Hot and Sour Chicken with Fungus

材料：
鸡腿1只、鸡翅膀2只、干木耳1小把、
红辣椒2个（切段）、姜4～5小片、葱1根

调味料：
（1）酒1大匙、酱油2大匙、醋1大匙、盐 1/3 茶匙、水 2/3 杯、淀粉水适量
（2）醋1大匙、麻油数滴

做法：
1. 鸡腿和鸡翅剁成小块。
2. 木耳泡至涨开，摘好，撕成小朵，洗净。
3. 用2大匙油爆香红辣椒段和姜片，放下鸡块同炒，炒至鸡块变色，淋下酒、酱油和醋后再炒一下。
4. 放下木耳、盐和水，炒匀后盖上锅盖，以中火焖煮5分钟。
5. 薄薄的勾一点芡，起锅时撒下葱段，再淋下调味料（2）的醋和麻油，大火炒匀即可。

Ingredients：
1 chicken leg, 2 chicken wings,
1 handful of dried black fungus,
2 red chilies (sectioned),
4～5 slices ginger, 1 stalk green onion

Seasonings：
(1) 1 tbsp wine, 2 tbsp soy sauce,
   1 tbsp vinegar, 1/3 tsp salt,
   2/3 cup water, a little of cornstarch paste
(2) 1 tbsp vinegar, a few drops of sesame oil

Procedures：
1. Cut chicken leg and wings into small pieces.
2. Soak dried fungus to soft, trim and tear to small pieces.
3. Stir-fry red chili sections and ginger slices with 2 tablespoons of oil, when fragrant, add chicken, stir-fry until the color of chicken turn light. Sprinkle wine, soy sauce and vinegar in, stir-fry again.
4. Add black fungus, salt and water, mix evenly and cook for 5 minutes over medium heat.
5. Thicken the sauce with cornstarch paste. Add green onion sections and then drizzle vinegar and sesame oil over chicken, mix and serve.

带骨鸡腿

泰式咖喱鸡
Curry Chicken, Thai Style

材料：
肉鸡鸡腿3只、马铃薯2个

调味料：
咖喱粉1 1/2 大匙、糖2茶匙、鱼露1大匙、
泰式红咖喱1大匙、辣油1大匙、清汤1 1/2 杯、椰浆1罐

做法：
1. 将每只鸡腿依大小剁成3或4块，汆烫一下，捞出。
2. 马铃薯去皮后切成大块。
3. 锅中先放1 1/2 大匙的油和咖喱粉，再开火慢慢将它们炒香。
4. 放入鸡腿、马铃薯和其余的调味料。
5. 用大火煮滚后转成小火，煮20～25分钟，至马铃薯已软时，即可关火，焖10分钟。
6. 再加热后装入深盘子或沙锅中上桌。

※煮时须不时搅动一下锅子，待马铃薯熟软后即可熄火，焖一下使它入味。
※泰式红咖喱买不到时可以用咖喱块1～2块来代替。

Ingredients:
3 chicken legs, 2 potatoes
Seasonings:
1 1/2 tbsp curry powder, 2 tsp sugar, 1 tbsp fish sauce,
1 tbsp red curry, 1 tbsp chili oil, 1 1/2 cups soup stock, 1 can coconut milk
Procedures:
1. Chop each chicken leg to 3 or 4 pieces, blanch and then rinse.
2. Peel and cut potato to large pieces.
3. Mix 1 1/2 tablespoons of oil with curry powder in a pot, turn on the heat, stir fry over low heat until fragrant.
4. Put chicken legs, potatoes and all the rest of seasonings in.
5. Bring to a boil over high heat, then reduce the heat to low, cook for 20 ~ 25 minutes until potatoes are soft. Turn off the heat, cover for 10 minutes.
6. Reheat the chicken, remove to a serving plate or a casserole.

※Keep on stirring the chicken while cooking to avoid from burning. Set aside for 10 minutes to make potato more tasteful.
※To make this dish, red curry is the best, otherwise curry cube can be substituted.

带骨鸡腿

韩式辣煮鸡
Spicy Chicken, Korean Style

材料：
鸡腿2只、鸡翅2只、大蒜5粒、干辣椒3个、姜末2茶匙、马铃薯400克、胡萝卜1小根、干香菇4朵、热水3杯、葱2根、洋葱 1/2 个

调味料：
麻油1大匙、酱油 1/4 杯、辣椒粉1大匙、韩式辣椒酱3大匙、胡椒粉 1/2 茶匙

做法：
1. 鸡腿和鸡翅分别剁成小块；大蒜切片；马铃薯、胡萝卜削皮、切块；香菇泡软、切片；干辣椒切小丁；葱切段；洋葱切粗丝。
2. 用麻油炒大蒜、姜末和干辣椒，爆香后放下鸡块炒至变色。
3. 鸡块变色后再放下香菇、马铃薯和胡萝卜，持续以大火来炒，加入调味料再炒，炒至香气透出，加入热水再煮滚。
4. 放下葱段和洋葱拌匀，改小火炖煮至鸡肉已熟，且汤汁煮剩下一半时即可。视鸡的品种，仿土鸡25～30分钟，肉鸡约20分钟。

Ingredients：
2 chicken legs, 2 chicken wings,
5 cloves garlic, 3 dried red chilies,
2 tsp chopped ginger, 400g. potatoes,
1 carrot, 4 pieces dried shitake mushroom,
3 cups hot water, 2 stalks green onion, 1/2 onion

Seasonings：
1 tbsp sesame oil, 1/4 cup soy sauce,
1 tbsp red chili powder, 3 tbsp Korean red chili paste, 1/2 tsp pepper

Procedures：
1. Chop chicken legs and wings to small pieces; slice garlic; peel tomato and carrot, then cut to pieces; soak shitake mushroom to soft, then slice it; dice dried red chili; cut green onion to sections; cut onion to wide stripes.
2. Heat sesame oil to fry garlic, ginger and dried chili until fragrant, add chicken in, continue to stir-fry.
3. When the color of the chicken has changed, add mushroom, potato and carrot in, stir-fry over high heat. Add seasonings in, stir-fry until fragrant. Add hot water in, bring to a boil.
4. Add green onion and onion in, stir-fry evenly. Turn to low heat, cook until chicken is cooked and the soup stock has half amount left. The cooking time is different, according to the type of chicken you use.

带骨鸡腿

意式鲜蔬烤鸡腿
Baked Chicken & Vegetables, Italian Style

材料:
棒棒鸡腿6只、西芹2棵、
洋葱1个、红甜椒1个

调味料:
盐1茶匙、胡椒粉 $1/2$ 茶匙、酒1大匙、
意大利综合香料 $1 \, 1/2$ 大匙、橄榄油1大匙

做法:
1. 鸡腿内侧切一道刀口,撒上盐、胡椒粉、酒和意大利综合香料,腌10~20分钟。
2. 蔬菜分别切适量,铺在烤盘内,再另外撒上适量的盐和1大匙的橄榄油拌匀。
3. 鸡腿皮面朝下,排放在蔬菜上。
4. 烤箱预热至220℃,放入鸡腿,烤15分钟后翻面,使鸡皮面朝上。
5. 再烤10~15分钟至鸡腿已熟、鸡皮金黄即可取出上桌。

Ingredients:
6 chicken drumsticks,
2 stalks celery, 1 onion, 1 red bell pepper

Seasonings:
1 tsp salt, $1/2$ tsp pepper, 1 tbsp wine,
$1 \, 1/2$ tsp Italian seasoning, 1 tbsp olive oil

Procedures:
1. Rinse and pat dry drum sticks. Score a cut on the meat side. Marinate with salt, pepper, wine and Italian seasoning for 10~20 minutes.
2. Cut all vegetables to wide strips, mix with salt and 1 tablespoon of olive oil, arrange on a baking ware.
3. Arrange drumsticks on top of vegetable with skin side down.
4. Preheat oven to 220℃, put chicken in bake for 15 minutes.
5. Turn chicken over, bake for another 10 ~ 15 minutes until chicken is done and the skin turn to golden brown. Remove and serve hot.

带骨鸡腿

蒜头焗鸡腿
Bake Chicken with Garlic

材料：
鸡腿2只、大蒜10粒、
洋葱 1/2 个、铝箔纸1大张、橄榄油1大匙

调味料：
酱油1大匙、糖 1/2 茶匙、
酒1大匙、盐 1/3 茶匙、胡椒粉 1/4 茶匙

做法：
1. 鸡腿剁成块，用调味料拌匀，腌20分钟。
2. 洋葱切成细条；大蒜切成厚片。
3. 把洋葱放在铝箔纸上，拌上少许橄榄油，再将鸡块放在上面，撒下大蒜，包好铝箔纸。
4. 烤箱预热至220～240℃，放入鸡腿烤20分钟，打开铝箔纸，再烤10分钟，至鸡腿表面焦黄，取出上桌。

Ingredients：
2 chicken legs, 10 cloves garlic,
1/2 onion, 1 piece foil, 1 tbsp olive oil

Seasonings：
1 tbsp soy sauce, 1/2 tsp sugar,
1 tbsp wine, 1/3 tsp salt, 1/4 tsp pepper

Procedures：
1. Cut each chicken into 4 or 5 pieces, mix with seasonings for 20 minutes.
2. Shred the onion; slice garlic.
3. Place onion on foil, mix with oil. Put chicken on top, then add garlic over chicken. Seal foil.
4. Preheat oven to 220～240℃, bake for 20 minutes. Unpack foil, bake for another 10 minutes to make the surface of chicken turn golden brown. Remove and serve hot.

带骨鸡腿

琥珀鸡冻
Jellied Chicken

材料：
鸡腿2只、鸡翅2只、猪皮200克（或鸡爪8只或用白明胶粉1包）、葱2根、姜2片、大料1粒

调味料：
酱油5大匙、酒2大匙、糖 1/2 大匙、盐 1/2 茶匙

做法：
1. 鸡腿和鸡翅连骨剁成小块；猪皮刮干净后切小块，和鸡腿一起用滚水烫2分钟，捞出洗净，放入汤锅中。
2. 加入葱、姜、大料及各种调味料与开水5杯，先用大火煮滚，立即改成小火，煮约30分钟。
3. 将鸡块先捡出，放在一个大碗内（或方的模型中）。
4. 猪皮捞出、切小一点，放回锅中，再以小火继续煮10分钟，使汤汁黏稠，剩下2杯。
5. 用细筛网将汤汁过滤到鸡块中，汤要盖过鸡肉，待凉透后，再放入冰箱内冷藏，4～5小时。
6. 食用时将凝固在表面的油脂刮除，以汤匙挖成小块，装入碟内。

※如用吉利丁粉（果冻粉），则用1杯冷水溶化之后，加入汤汁中，煮滚即可。

Ingredients:
2 chicken legs, 2 chicken wings, 200g. pork skin (or 8 chicken feet, or 1 pack unflavored gelatin), 2 stalks green onion, 2 slices ginger, 1 star anise

Seasonings:
5 tbsp soy sauce, 2 tbsp wine, 1/2 tbsp sugar, 1/2 tsp salt

Procedures:
1. Clean and chop the chicken into small pieces. Cut pork skin into pieces too. Blanch all for 2 minutes. Drain and rinse them, then place in a pot.
2. Add green onion, ginger, star anise, all the seasonings, and 5 cups of boiling water. Bring to a boil. Turn to low heat, simmer for 30 minutes.
3. Remove chicken and arrange in a bowl.
4. Chop the pork skin and return to the pot. Simmer for another 10 minutes until the soup becomes thicker (there should be about 2 cups of soup left).
5. Strain the soup into the chicken. When it cools, store in the refrigerator for about 4~5 hours, until the soup is firmed.
6. Skin off grease from surface after it firmed. Scoop the chicken, and transfer onto a serving plate.

※If you use unflavored gelatin, dissolve in 1 cup of cold water and then add to chicken soup in #3 instead of #4. Bring to a boil.

带骨鸡腿

洋葱烧鸡
Stewed Chicken with Onion

材料：
仿土鸡鸡腿2只、鸡翅2只、
洋葱1个、红葱4粒、洋菇8～10朵
调味料：
淡色酱油3大匙、酒2大匙、
醋1大匙、冰糖1茶匙

做法：
1. 鸡腿和鸡翅分别剁成块；洋葱切宽条；红葱头切片；洋菇大的切对半，小的不切。
2. 锅中烧热2大匙油炒鸡块，炒至鸡肉变色后盛出。
3. 另加入1大匙油炒红葱片和洋葱，待香气透出时，放回鸡块和洋菇，再炒一下。
4. 加入调味料，炒煮约1分钟，加入2杯半的热水，再煮滚后改成小火，炖烧50～60分钟，至汤汁浓稠即可。

※喜欢吃口感较嫩的鸡肉的话，只要烧约20分钟，鸡肉熟了即可。

Ingredients:
2 simulate native chicken legs,
2 chicken wings,
1 medium sized onion,
4 cloves shallot,
8～10 pieces mushroom

Seasonings:
3 tbsp light colored soy sauce,
2 tbsp wine, 1 tbsp vinegar,
1 tsp rock sugar

Procedures:
1. Chop chicken legs and wings to pieces; cut onion to wide strips; slice shallots; halve mushroom.
2. Heat 2 tablespoons of oil to stir-fry chicken, remove when the color turn light.
3. Add 1 tablespoon of oil to sauté onion and shallot until fragrant. Return chicken and mushrooms back, continue to stir-fry for a while.
4. Add all seasonings in, boil for 1 minute. Add 2 1/2 cups of hot water in, reduce to low heat when it boils again. Stew for about 50~60 minutes until the tenderness you like, and the sauce is thickened.

※The texture of chicken is quite different if you stew chicken to just cooked only for about 20 minutes. You may try it!

带骨鸡腿

梅酱鸡
Chicken With Plum Sauce

材料：
鸡腿2只、葱2只、姜2片、大料 1/2 粒、陈皮2片

调味料：
酱油6大匙、酒2大匙、糖 1/2 大匙

梅酱料：
话梅6粒、苏州梅10粒、冰糖2大匙、白醋2大匙

做法：
1. 鸡腿洗净、擦干水分，用约2大匙酱油涂在表皮，并泡5分钟。
2. 将1杯油烧得很热，鸡皮朝下、把鸡腿放入油中，炸黄表皮，取出。
3. 用1大匙油爆香葱段和姜片，并加入剩余的4大匙酱油、调味料、大料及陈皮，注入开水2杯，放入鸡腿，以小火煮20分钟，浸泡至凉后取出。
4. 苏州梅与话梅加水2杯，蒸20分钟，取出待凉后剥下梅肉，连汁放入小锅中。
5. 加入冰糖、白醋和 1/2 杯煮鸡汁，用小火再煮10分钟，当梅酱浓稠时盛出，装在碗中。
6. 将鸡腿切成宽条、装碟，浇上适量的梅子酱供食。

※余下的梅子酱，可装瓶存放冰箱中留用。

Ingredients:
2 chicken legs, 2 stalks green onion, 2 slices ginger, 1/2 star anise, 2 small pieces dried orange peel

Seasonings:
6 tbsp soy sauce, 2 tbsp wine, 1/2 tbsp rock sugar

Plum sauce:
6 dried plums, 10 preserved plums, 2 tbsp rock sugar, 2 tbsp white vinegar

Procedures:
1. Clean and pat the legs dry. Soak in 2 tablespoons of soy sauce for 5 minutes.
2. Heat 1 cup of oil, put chicken in with skin side down. Fry until the skin side become golden brown. Drain.
3. Heat 1 tablespoon of oil to sauté green onions and ginger. Add the seasonings, star anise, dried orange peel, and 2 cups of hot water. Add chicken legs in, cook over low heat for about 20 minutes. Remove chicken after cools.
4. Steam the two kinds of plums with 2 cups of hot water for 20 minutes. Remove and let it cools. Discard the kernel and keep the pulp.
5. Mix pulp with rock sugar, vinegar, and 1/2 cup of chicken broth, cook for about 10 minutes until the plum sauce become thicker. Remove.
6. Chop chicken legs into wide pieces, place on a serving plate and then pour the plum sauce over the meat. Serve.

※Keep the plum sauce in a bottle and store in fridge for later use.

带骨鸡腿

Stewed Chicken with Mushroom & Bamboo Shoot

红烧香菇竹笋鸡

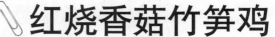

Stewed Chicken with Mushroom & Bamboo Shoot

材料：
半土鸡腿2只、花菇5朵、
竹笋450克、葱2根、姜2片、香菜少许

调味料：
酒2大匙、酱油2大匙、热水2杯、冰糖1茶匙、盐适量、淀粉水适量

做法：
1. 鸡腿剁成块；香菇泡软、切片；竹笋剥壳，切成滚刀块；葱切长段。
2. 锅中烧热2大匙油，放下鸡块炒至鸡肉变色，血水封住。
3. 再加入香菇、竹笋块、葱段和姜片一起炒香，淋下酒、酱油和冰糖。
4. 再炒两三下即可注入热水，煮滚后改成小火，烧1小时以上，至喜爱的烂度，或只烧煮25～30分钟以保持较嫩的口感。
5. 适量加盐调味。如汤汁仍多，可以用淀粉水略勾薄芡。装盘后饰以香菜。

※笋子容易吃进咸味，因此这种红烧鸡的酱油不要多、不要太咸，以鸡的鲜味搭配花菇及竹笋的香气。
※用栗子烧鸡也很好吃，栗子去壳后用热水烫一下，浸泡5～10分钟，剥去薄膜，蒸熟后再和鸡烧一下即可。

Ingredients:
2 simulate native chicken legs, 5 dried shitake mushrooms, 450g. bamboo shoot, 2 stalks green onion, 2 slices ginger, cilantro leaves for decoration

Seasonings:
2 tbsp wine, 2 tbsp soy sauce, 2 cups hot water, 1 tsp rock sugar, salt to taste, cornstarch paste

Procedures:
1. Cut the chicken to pieces; soak and then slice mushrooms; trim and cut bamboo shoot to pieces; cut green onions to long sections.
2. Heat 2 tablespoons of oil to stir-fry chicken until the color has changed.
3. Add mushroom, bamboo shoot, green onion and ginger slices in, stir-fry until fragrant. Add wine, soy sauce and rock sugar.
4. Stir evenly, add hot water in, bring to a boil, simmer for about 1 hour until the tenderness you like or you may just stew for 25~30 minutes to keep it's chewy texture.
5. Season with salt if needed. You may thicken the soup with some cornstarch paste. Decorate with cilantro.

※Bamboo shoots turn salty easily; therefore do not add too much soy sauce when making this dish. The key is to use the natural flavor of shitake mushroom and bamboo shoots to enhance the flavor of chicken.
※Chestnut is also a good ingredient to add in. After peeling off the shell, add chestnuts in hot water and soak for 5-10 minutes. Peel off the membranes, steam until done, and then cook the chestnuts with chicken for a few minutes.

带骨鸡腿

Deep-fried Chicken with Vegetable Rice

炸鸡腿菜饭
Deep-fried Chicken with Vegetable Rice

材料（2人份）：
鸡腿2只、葱1根、大蒜2粒、姜2片、大料 1/2 粒、小油菜200克、米2杯
调味料：
（1）酱油3大匙、糖1茶匙、酒1大匙、水3大匙
（2）盐 1/4 茶匙、胡椒粉适量

做法：
1. 在鸡腿内侧顺着大骨，用刀子切割一道刀口，并用叉子叉些洞孔。
2. 将调味料（1）调匀，放入鸡腿和拍碎的葱段、姜片、大蒜和大料，一起腌1小时以上。
3. 把4杯油烧至7分热，放入鸡腿，以中小火炸熟（6～7分钟），取出。
4. 油再烧热，放下鸡腿，大火再炸30秒钟，至外皮酥脆，捞出，盛放在菜饭，撒上胡椒粉即可。

菜饭：
小油菜洗净，切成3厘米长的段，锅中烧热2大匙油来炒小油菜，炒至微软后加少许盐调味。放入洗好的米中，略拌均匀后加入1 1/2 杯水，放入电锅中煮成饭，便是菜饭。

Ingredients 〈2 servings〉:
2 chicken legs, 1 stalk green onion, 2 cloves garlic,
2 slices ginger, 1/2 star anise, 300g. green cabbage, 2 cups rice

Seasonings:
(1) 3 tbsp soy sauce, 1 tsp sugar, 1 tbsp wine, 3 tbsp water
(2) 1/4 tsp salt, pepper to taste

Procedures:
1. Make a deep cut along the bone on the meat side of chicken leg. Pierce the meat with a fork.
2. Mix seasonings (1) with chicken legs, crushed green onion, ginger, garlic, and star anise, marinate for at least 1 hour.
3. Heat 4 cups of oil to 140℃, deep-fry chicken for about 6～7 minutes over medium-low heat until done. Remove chicken.
4. Reheat oil to very hot, deep-fry chicken again over high heat for 30 seconds. Remove when the skin is crispy, serve with cooked rice, sprinkle some pepper.

Cooked rice with green cabbage:
Rinse and trim green cabbage, cut to 3 cm long. Stir-fry green cabbage with 2 tablespoons of oil, season with salt. Mix with rinsed rice, add 1 1/2 cups of water in. Cook in a rice cooker until cooked.

细说鸡翅
All about Chicken Wings

　　鸡翅膀是属于活动较多的部分，因此肉质细嫩。鸡翅因为肉质不多，常是整只红烧、卤煮、炸或烤。较费工夫的就是去除两只翅骨后再做变化的菜式。

　　鸡翅可以分为翅根（又称小鸡腿）、中翅和翅尖三个部位（图1）。翅根因为接近鸡胸，肉质较干硬，而翅尖又没有肉，因此最常用的就是中间的翅膀部分，把翅根和翅尖放在一起烧，则可以增加分量。

The wings of chicken are very active; therefore the meat is delicate and tender. Since there is not much meat in the wings, they are generally cooked in whole and are suitable for cooking methods such as stewing, deep-frying or baking. The more labor intensive way to prepare a dish is to separate the meat from the wing bone prior to using it in the dish.

There are three parts in a chicken wing: drummette, wingette, and tip (pic 1). The drummette is adjacent to the breast; therefore the meat is dryer and harder. The tip has no meat. As a result, wingette becomes the most used part of a wing. You may increase the portion by including the drummettes and tips.

图1

辣烤鸡翅
Spicy Wings

材料:
鸡翅膀10只、洋葱 1/2 个、
大蒜2粒、辣椒粉1茶匙

调味料:
酱油 1/2 大匙、酒1大匙、
盐 1/2 茶匙、辣椒粉1茶匙

做法:
1. 洋葱切丝、大蒜拍碎,一起和调味料混合拌匀成腌料。
2. 放入鸡翅和腌料拌匀,腌20分钟。
3. 烤箱预热至220℃,放入鸡翅,先将鸡翅皮面朝下,烤8分钟。
4. 翻面后再涂上一层腌料,再烤10分钟,取出,趁热再撒一些辣椒粉。

Ingredients:
10 chicken wings, 1/2 onion,
2 cloves garlic, 1 tsp chili powder

Seasonings:
1/2 tbsp soy sauce, 1 tbsp wine,
1/2 tsp salt, 1 tsp chili powder

Procedures:
1. Shred onion; crush garlic, mix together with seasonings.
2. Mix chicken wings with seasonings, marinate for 20 minutes.
3. Preheat oven to 220℃, bake wings with skin side down for 8 minutes.
4. Remove wings, turn it over and brush marinade once more, bake for 10 minutes more. Remove, sprinkle some chili powder while it is hot.

龙凤串翅
Stuffed Chicken Wings with Ham

材料：
鸡翅膀10只、金华火腿60克、
笋1根、小油菜250克、清汤 1/3 杯

酒2茶匙、盐 1/3 茶匙、清汤5大匙、淀粉水 1/2 茶匙

1. 将鸡翅膀的翅尖剁下（只用整只鸡翅），再将鸡翅前端关节处也剁掉1厘米，（使翅膀中的两只骨头露出一点，如图1所示），全部在开水中烫煮约半分钟，捞出、冲一下冷水。
2. 火腿蒸熟；笋煮熟。将火腿及笋分别切成约4厘米长的粗条。
3. 从烫熟的鸡翅内抽出2只骨头（图2），在其空洞处塞入1只火腿及1根笋条。
4. 把鸡翅排在蒸碗内，要鸡皮面朝下，淋下酒、盐及清汤，放入锅中，用大火蒸约20分钟。
5. 蒸好后，先把碗内的汤汁倒入小锅中，另加 1/3 杯清汤一起煮滚，用少许淀粉水勾成薄芡，淋到扣在盘中的鸡翅上面，盘边围放炒过的小油菜。

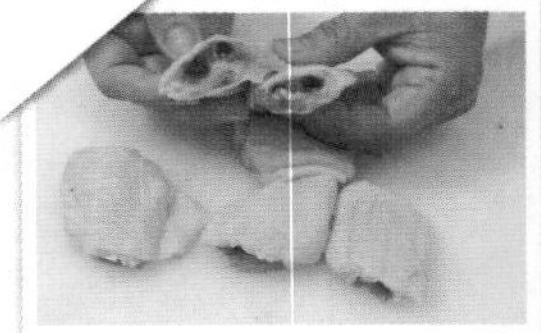

图1

图2

Ingredients:
10 chicken wings, 60g. Chinese ham, 1 bamboo shoot, 250g. green cabbage, 1/3 cup soup stock

Seasonings:
2 tsp wine, 1/3 tsp salt, 5 tbsp soup stock, 1/2 tsp cornstarch paste

Procedures:
1. Cut off the tip of chicken wing, then remove about 1 cm from each side of the wing. Cook wings in boiling water for about 1/2 minute. Remove and plunge into cold water.
2. Steam Chinese ham to done; cook bamboo shoot. Cut two kinds of ingredients into match-stick-size. 10 sticks for each.
3. Push the meat of chicken wings to remove the two bones from chicken wing. Stuff bone holes with 1 stick of ham and 1 stick of bamboo shoot.
4. Arrange stuffed chicken wings in a bowl (the skin side down), add wine, salt, and soup stock in. Steam over high heat for 20 minutes.
5. Pour stock from the bowl into a sauce pan, add 1/3 cup of soup stock, bring to a boil and thicken with cornstarch paste. Turn the bowl upside-down and place the chicken wings on the serving plate. Pour the thickened sauce over the wings. Serve with stir-fried vegetable.

杏鲍凤翼煲
Chicken Wings and Mushrooms in Casserole

材料：
鸡翅6只、杏鲍菇4朵、西生菜 1/3 个、
葱2根、姜片3～4片

调味料：
酒1大匙、酱油1大匙、蚝油1大匙、糖 1/2 茶匙、水1 1/2 杯

做法：
1. 把鸡翅的翅膀和翅尖部分剁开；杏鲍菇切成厚片；西生菜撕成大片；葱切成长段。
2. 起油锅用2大匙油爆香葱段和姜片，再放下鸡翅，煎黄鸡皮表面，淋下酒、酱油和蚝油同炒，炒到鸡翅上色。
3. 加入杏鲍菇和翅尖炒一下，淋下水、加入糖，以小火烧煮20分钟，至汤汁剩下半杯左右。
4. 西生菜在滚水中烫一下，放入沙锅中垫底，倒入鸡翅，再煮滚即可。

Ingredients:
6 chicken wings, 4 pieces xing-bao mushroom,
1/3 head lettuce, 2 stalks green onion, 3 ~ 4 slices ginger

Seasonings:
1 tbsp wine, 1 tbsp soy sauce,
1 tbsp oyster sauce, 1/2 tsp sugar, 1 1/2 cups water

Procedures:
1. Separate chicken wing and tip; slice xing-bao mushroom; tear lettuce to large pieces; cut green onion to long sections.
2. Sauté green onion and ginger with 2 tablespoons of oil. Add chicken wings in, fry until the skin become light browned. Add wine, soy sauce and oyster sauce in, stir-fry until wings get browned.
3. Add mushrooms and wing tips, stir-fry for a while. Add water and sugar in, cook over low heat for 20 minutes until the water reduced to half a cup.
4. Blanch lettuce, place in a casserole, pour all wings in, bring to a boil again. Turn off the heat, serve hot.

玉米烩鸡翅
Stewed Chicken Wings with Corn

材料：
鸡翅8只、葱1根、姜1片、胡萝卜1根、
玉米粒1杯、洋葱丝 1/2 杯、面粉1～2大匙

调味料：
酒1大匙、盐 1/2 茶匙、水2杯、奶油1大匙

做法：
1. 在3杯水中加葱1根和姜1片煮滚，放下鸡翅烫2分钟，捞出；胡萝卜切条。
2. 用2大匙油爆香洋葱，放下鸡翅拌炒，淋酒烹香后再加入水和盐。煮10分钟后，加入胡萝卜再煮5～6分钟。
3. 放下玉米粒再煮一滚，用筛网将面粉筛入汤汁中（视汤汁的多少决定加面粉的量，要慢慢地加入），边筛边搅匀，成浓稠状，以小火再煮1分钟。
4. 再调一下味道，加入奶油搅匀以增加香气。

※这是一道口味清淡的菜，因此鸡翅不炸而是烫煮。

Ingredients:
8 chicken wings, 1 stalk green onion,
1 slice ginger, 1 carrot,
1 cup sweet corn kernels,
1/2 cup onion shreds, 1 ~ 2 tbsp flour

Seasonings::
1 tbsp wine, 1/2 tsp salt, 2 cups water, 1 tbsp butter

Procedures:
1. Boil green onion and ginger with 3 cups of water, put chicken wings in, blanch for 2 minutes, drain. Cut carrot into stripes.
2. Sauté onion with 2 tablespoons of oil, then add chicken wings in, stir-fry for a while. Add wine, water, and salt, cook for 10 minutes. Put carrot in, cook for another 5 ~ 6 minutes.
3. Add sweet corn, bring to a boil. Sieve flour into soup little by little, stir the soup while sieving, sieve until soup is thickened. Simmer for 1 minute.
4. Season again if needed, add butter at last to enhance the flavor.

※Be sure to boil the wings first to clean it and remove the bad flavor.

香辣凤翼
Spicy Chicken Wings

材料：
三节鸡翅膀5只、干辣椒1杯、
葱2根、姜末1茶匙、大蒜末1大匙、花椒粉1茶匙、炸油4杯

腌鸡料：
葱1根（拍碎）、姜2片（拍碎）、酱油1大匙、酒1大匙、盐 1/4 茶匙、糖1茶匙、花椒粉 1/4 茶匙

做法：
1. 鸡翅膀洗净，擦干水分，剁成小块，用腌鸡料拌匀，腌15～20分钟。
2. 干辣椒只要用湿纸巾擦一下；葱切成葱花。
3. 锅中的炸油烧到8分热，放下鸡翅，用中火炸至熟且香，捞出，油倒出。
4. 用2大匙油把干辣椒快速炒红，盛出。加入姜、蒜末和鸡翅膀，一起翻炒均匀。
5. 关火，撒下葱花和花椒粉，放回干辣椒，再拌均匀、装盘。

※鸡翅膀也可以炸两次，使它更加香酥

Ingredients：
5 chicken wings, 1 cup dried red chilies, 2 stalks green onion,
1 tsp chopped ginger, 1 tbsp chopped garlic, 1 tsp brown peppercorn powder,
4 cups of oil for deep-fry

Seasonings：
1 stalk green onion (crushed), 2 slices ginger (crushed), 1 tbsp soy sauce,
1 tbsp wine, 1/4 tsp salt, 1 tsp sugar, 1/4 tsp brown pepper powder

Procedures：
1. Rinse and pat dry the wings, chop into small pieces, mix with seasonings, marinate for 15～20 minutes.
2. Wipe dried chilies with wet paper towel; Chop green onion.
3. Heat oil to 160℃, deep-fry wings over medium heat until done, drain. Pour away oil.
4. Stir fry dried red chilies with 2 tablespoons of oil until it turns bright red, remove. Add ginger, garlic and chicken wings, stir fry evenly.
5. Turn off the heat, sprinkle green onion and brown pepper powder in, return dried red chilies, mix well, remove and serve.

※You may deep-fry chicken wings two times to make it more crispy and fragrant.

笋烧双宝
Stewed Two Treasures with Bamboo Shoot

材料:
鸡翅膀8只、鸡肫8个、笋2棵、葱2根、姜2片、香菜少许

调味料:
酱油3大匙、酒1大匙、糖 1/2 大匙、大料半粒、水2 1/2 杯、盐适量

做法:
1. 将翅膀和翅尖剁开，用滚水汆烫1分钟，捞出、洗净。
2. 鸡肫摘洗干净、烫过、沥干；笋削皮，切成滚刀块或厚片。
3. 用1大匙油爆香葱段、姜片和大料，淋下酒，再加酱油、糖和水，煮滚。
4. 把鸡翅、鸡肫和笋块放入锅中，再煮滚后改小火，烧20～25分钟。
5. 尝一下味道后略勾薄芡，盛入盘中，可以点缀些香菜。

Ingredients:
8 chicken wings,
8 gizzards, 2 bamboo shoots,
2 stalks green onion,
2 slices ginger, cilantro leaves

Seasonings:
3 tbsp soy sauce, 1 tbsp wine,
1/2 tbsp rock sugar, 1/2 star anise,
2 1/2 cups water, salt to taste

Procedures:
1. Cut wing tip off from chicken wing, blanch for 1 minute. Drain and rinse it.
2. Clean gizzards, blanch and drain. Trim bamboo shoot, cut to thick pieces.
3. Heat 1 tablespoon of oil to stir-fry green onion, ginger and star anise until fragrant, sprinkle wine in, then add soy sauce, sugar and water, bring to a boil.
4. Add chicken wings, gizzards, and bamboo shoot, reduce to low heat after boils again. Simmer for 20 ~ 25 minutes.
5. Season again if needed. Thicken with cornstarch paste, remove to plate, decorate with cilantro.

香柠焖鸡翅
Stewed Wings with Lemon

材料：
鸡翅膀10只、柠檬1个、红葱头3粒、洋葱 1/3 个、大蒜2粒

调味料：
酱油2大匙、酒1大匙、冰糖1大匙、胡椒粉 1/4 茶匙

做法：
1. 鸡翅膀和翅尖剁开，洗净、擦干。
2. 红葱头切片；大蒜切片；洋葱切丝；柠檬挤汁，约有2大匙备用。
3. 锅中热1大匙油，放下鸡翅，鸡皮面朝下煎一下，煎至略有焦痕。
4. 放下洋葱丝、红葱片和大蒜片一起炒香，再加入所有调味料和柠檬汁一起煮滚，同时不断地炒煮，至鸡翅上色。
5. 加入约1杯半的水，以小火煮20分钟至鸡翅已熟，且汤汁浓稠即可关火。

Ingredients:
10 chicken wings, 1 lemon, 3 shallots, 1/3 onion, 2 cloves garlic

Seasonings:
2 tbsp soy sauce, 1 tbsp wine, 1 tbsp rock sugar, 1/4 tsp pepper

Procedures:
1. Separate chicken wing and tip, rinse and pat dry.
2. Slice shallot and garlic; shred onion; juiced the lemon, about 2 tablespoons.
3. Heat 1 tablespoon of oil to fry chicken wings until light browned.
4. Add onion shreds, shallot slice and garlic slice, stir-fry until fragrant. Add seasonings and lemon juice, stir-fry until wings are colored.
5. Add 1 1/2 cups of water in, simmer for about 20 minutes until sauce is reduced and thickened. Turn off the heat.

细说鸡爪
All about Chicken Feet

鸡爪又称为"凤爪",没有肉仅有皮,是含有丰富胶质的部分,可以补筋骨,近几年又被视为可以美容养颜的胶原蛋白,因此价格上扬。

鸡爪的烹调方法不多,多半是炖汤、卤或红烧,广东饮茶点心中的"豉汁蒸凤爪"和凉拌菜中的"去骨凤爪"是较有名的菜式。鸡爪中的胶质也被用来做"冻菜"或做汤包时不可少的"皮冻"的原料。

Chicken feet are also called phoenix talons. They have only skin and no meat. However, chicken feet have high collagen content and are also gelatinous. Some people also claim that chicken feet can prevent or treat wrinkles due to the high collagen content. Chicken feet are mostly stewed, braised or making soup The gelatin content in chicken feet is one of the ingredients for jello dishes.

Ingredients:
10 chicken feet,
1 chicken breast,
5 shitake mushrooms,
1 tbsp sweet almonds,
1 tsp bitter almonds, 2 slices ginger

Seasonings:
1 tbsp wine, salt to taste

Procedures:
1. Cut each chicken foot to two pieces, blanch together with chicken breast for 1 minute. Remove and rinse.
2. Soak shitake mushrooms to soft, cut off the stems, mix with cornstarch to clean mushrooms. Rinse and then slice it.
3. Bring 8 cups of water to a boil, add chicken feet, chicken breast, mushrooms, almonds, and ginger in, then add wine. Bring to a boil over high heat, then reduce the heat to low.
4. Simmer for 1 hour until feet are tender enough, season with salt.

北菇炖凤爪汤
Chicken Feet Soup with Shitake Mushrooms

材料:
鸡爪10只、鸡胸1个、香菇5朵、南杏1大匙、北杏1茶匙、姜2片

调味料:
酒1大匙、盐适量调味

做法:
1. 将每只鸡爪剁成两段,和鸡胸肉一起用开水烫1分钟。捞出后洗净。
2. 香菇用冷水泡软,剪下菇蒂后用干淀粉抓洗一下,再以水冲洗一下,改刀切片。
3. 烧滚8杯水,放入鸡爪、鸡胸肉、香菇、南杏、北杏和姜片,淋下酒,先以大火煮滚,再改小火慢炖。
4. 鸡爪炖约1小时至够软烂,加适量的盐调味。

豉汁凤爪

Steamed Chicken Feet with Black Bean Sauce

材料：
鸡爪10只、盐2茶匙、面粉2大匙、酱油2茶匙、淀粉2茶匙、红辣椒末2大匙

豉汁料：
豆豉2大匙、大蒜末2茶匙、红葱末2茶匙、酒1大匙、蚝油1大匙、糖2茶匙、淀粉水适量

做法：
1. 鸡脚剪去爪尖，剁成两段。用约2茶匙的盐抓洗一下，冲洗干净后再加入面粉搓洗一下，冲洗干净，擦干。
2. 鸡爪加酱油拌匀，放置10分钟。烧热1杯油，分次来炸鸡爪，炸黄之后捞出，拌上少许淀粉，盛放在深盘中。
3. 豆豉加1/3杯水泡15分钟，取出豆豉，汤汁留用。
4. 用1大匙油炒香豉汁料中的大蒜和红葱末，再放下豆豉，以小火继续炒。淋下酒和蚝油，再加入糖和泡豆豉的水，小火煮3分钟。用适量淀粉水勾成浓芡，盛出，铺放在鸡爪上。
5. 上锅蒸30分钟，撒上红椒末，再蒸10分钟

Ingredients:
10 chicken feet, 2 tsp salt, 2 tbsp flour,
2 tsp soy sauce, 2 tsp cornstarch,
2 tbsp chopped red chili

Seasonings:
2 tbsp fermented black beans,
2 tsp chopped garlic,
2 tsp chopped shallot, 1 tbsp wine,
1 tbsp oyster sauce, 2 tsp sugar,
cornstarch paste

Procedures:
1. Trim off nail from chicken feet, then cut into two parts. Clean feet with 2 teaspoons of salt, rinse and then mix with flour, rinse again, drain and pat dry.
2. Mix feet with soy sauce, set aside for 10 minutes. Heat 1 cup of oil to deep-fry chicken feet until brown. Mix with a little of cornstarch. Place on a steaming plate.
3. Soak fermented black beans with 1/3 cup of water for 15 minutes. Remove beans, keep water for later use.
4. Sauté garlic and shallot with 1 tablespoon of oil, add black beans, fry over low heat for a while. Add wine, oyster sauce, sugar and reserved water, simmer for 3 minutes. Thicken with cornstarch paste, remove and spread over chicken feet.
5. Steam for 30 minutes, sprinkle red chili over, continue to steam for 10 minutes.

细说鸡翅

图书在版编目（CIP）数据

家家锅中有只鸡/程安琪著．—北京：中国农业大学出版社，2009.8
ISBN 978-7-81117-653-7

Ⅰ．家… Ⅱ．程… Ⅲ．鸡肉-菜谱 Ⅳ．TS972.125

中国版本图书馆CIP数据核字（2008）第205221号

> 中文简体字版©2008年由中国农业大学出版社出版发行，本书经由台湾旗林文化出版有限公司独家授权出版，非经书面同意，不得以任何形式再利用。

著作权合同登记图字：01-2008-5076

书　　名	家家锅中有只鸡	
作　　者	程安琪　著	
策划编辑	六睿工社	责任编辑　周　伟
装帧设计	东方黑马	责任校对　王晓凤　陈　莹
出版发行	中国农业大学出版社	
社　　址	北京市海淀区圆明园西路2号	邮政编码　100193
电　　话	发行部 010-62731190,2620	读者服务部 010-62732336
	编辑部 010-62732617,2618	出 版 部 010-62733440
网　　址	http://www.cau.edu.cn/caup	E-mail　cbsszs@cau.edu.cn
经　　销	新华书店	
印　　刷	涿州市星河印刷有限公司	
版　　次	2009年8月第1版　2009年8月第1次印刷	
规　　格	889×1 194　24开本　7印张　190千字	
定　　价	19.80元	

图书如有质量问题本社发行部负责调换